TK 2511 .C5

Clifford, Martin,

Electric/electronic motor
data handbook

Electric/Electronic Motor Data Handbook

MARTIN CLIFFORD

PRENTICE HALL, Englewood Cliffs, New Jersey 07632

Library of Congress Cataloging-in-Publication Data

CLIFFORD, MARTIN
  Electric/electronic motor data handbook / Martin Clifford.
    p. cm.
  ISBN 0-13-199696-7
  1. Electric motors—Handbooks, manuals, etc. 2. Electric
motors—Electronic control—Handbooks, manuals, etc. I. Title.
TK2511.C55 1990
621.46—dc20   89-26618
                                                            CIP

Editorial/production supervision
  and interior design: Rob DeGeorge
Cover design: Wanda Lubelska
Manufacturing buyer: Ray Sintel

 © 1990 by Prentice-Hall, Inc.
A Division of Simon & Schuster
Englewood Cliffs, New Jersey 07632

The publisher offers discounts on this book when ordered
in bulk quantities. For more information, write:

Special Sales/College Marketing
College Technical and Reference Division
Prentice Hall
Englewood Cliffs, New Jersey 07632

All rights reserved. No part of this book may be
reproduced, in any form or by any means,
without permission in writing from the publisher.

Printed in the United States of America

10 9 8 7 6 5 4 3 2 1

ISBN 0-13-199696-7

PRENTICE-HALL INTERNATIONAL (UK) LIMITED, *London*
PRENTICE-HALL OF AUSTRALIA PTY. LIMITED, *Sydney*
PRENTICE-HALL CANADA INC., *Toronto*
PRENTICE-HALL HISPANOAMERICANA, S.A., *Mexico*
PRENTICE-HALL OF INDIA PRIVATE LIMITED, *New Delhi*
PRENTICE-HALL OF JAPAN, INC., *Tokyo*
SIMON & SCHUSTER ASIA PTE. LTD., *Singapore*
EDITORA PRENTICE-HALL DO BRASIL, LTDA., *Rio de Janeiro*

**To Adrienne—
My Alter Ego
On A Higher Level**

Contents

PREFACE *xv*

ACKNOWLEDGMENTS *xvii*

1   BASIC DATA   1

Voltage *1*
  Units of Voltage—The Volt *1*
Voltage Conversions *2*
  The Microvolt *3*
  The Millivolt *3*
  The Kilovolt *3*
  The Megavolt *4*
Abbreviations *4*
Source Voltage *4*
  DC Voltage Symbols *4*
Current *5*
  The Milliampere *5*
  The Microampere *6*
  Current Conversion Rules *6*
  Current Designations *7*
Resistance *8*

The Kilohm *8*
The Megohm *8*
Resistor Power Designations *9*
Resistor Types *9*
Codes and Values *12*
  Resistor Symbols, 12 Resistor Color Code, 12
Tolerance *14*
Resistor Connections *14*
  Resistors in Series *15*
  Two Resistors in Parallel *15*
  The 10 to 1 Rule *16*
  Three or More Resistors in Parallel *16*
  Resistors in Series-Parallel *17*
  Summary of Resistor Connections *20*
Resistivity of Metals *21*
Conductivity *21*
  Summary of Conductances *22*
Insulators *23*
Voltage, Current, and Resistance Relationships *23*
  Summary of Ohm's Law for DC *24*
IR Drops *25*
  Polarity of IR Drops *25*
  Summary of Parallel and Series Voltage Drops *26*
Loading *27*
Power in DC Circuits *27*
  Power-Line Loss *28*
DC Power Laws *28*
  Multiples of the Watt *29*
  Total Power Dissipation *32*
  Summary of Power Relationships *32*
Work *33*
  Units of Work *34*
  The Time Element *34*
Power versus Energy *35*
Heating Effect of a Current *35*
Nomograms *35*
Temperature *38*
  Temperature Conversions *39*
  Temperature Coefficient of Resistance *41*
Wire *42*
  Cross-Sectional Area of Wire *44*
  Resistance of Wire *44*
  Fusing Currents of Wires *46*
  Current versus Wire Size *47*

Contents   vii

      Fuses for Motors *47*
         Fuse Types *47*
         Fuse Ratings *49*
         The Plug Fuse *49*
         Cartridge Fuses *50*
         Fuse Pullers *50*
         Circuit Breakers *51*
         Plug-to-Circuit-Breaker Conversions *52*
      Horsepower *53*
         Summary of Formulas Involving Horsepower *56*
      The Foot-Pound *56*
      Metric and English Units *58*
      Torque *60*
         Pound-Feet *60*
      Efficiency *62*
         Summary of Efficiency Formulas *63*
         Overall Efficiency *63*
      Relays *65*

2 DC POWER SOURCES *67*

      Batteries *67*
         Physical Dimensions *67*
         Cells in Series *68*
         Cells in Parallel *69*
         Cells in Series-Parallel *69*
         Cell Polarity *70*
         Cells in Series Opposing *70*
      Ampere-Hour Ratings *70*
      Reference Points *71*
         Voltage Reference Points *71*
      Battery Types and their Characteristics *71*
         Battery Identification *72*
         Lead-Acid Batteries *73*
         Nickel-Cadmium Cells *74*
         Zinc-Carbon Cells *75*
         Alkaline Cells *78*
         Mercury Cells *78*
         Lithium Cells *81*
         Silver Oxide Cells *82*
         Zinc Chloride Cells *83*
         Zinc-Air Cells *84*
      Internal Resistance of a Battery *84*

Battery Efficiency 85
Electronic Power Supplies 86
  Dual-Polarity Power Supplies 87
  Determination of the Charge Rate 89
  Voltage Regulation 89
Voltage Regulators 89
Surge Limiting 90
Zener-Diode Regulator 90
  Series Zeners 91
    Cascaded Zener Regulation, 92 Low-Voltage Motor Control, 92
  Diode Voltage Control 92
  Transistor Voltage Regulators 93
  The Series Zener and Transistor Regulator 94
  Regulation with a Differential Amplifier 94
  Shunt Transistor Voltage Control 94
  Darlington Motor Control 95
  The Shunt Regulator 96
Motor Generators 96
  Advantages and Disadvantages of the Motor Generator 97
The Dynamotor 98
The Rotary Converter 99

3 AC FOR MOTORS *101*

Waveform Variation *101*
Voltage and Current Measurements *103*
The AC Sine Wave *103*
  Average Values *104*
  Instantaneous Values of Sine Voltages and Currents *105*
  Effective Values of Sine Voltages and Currents *106*
  The Variable RMS *108*
  Conversions *109*
  Summary of Sine-Wave Voltage and Current Formulas *109*
Phase *110*
  Generator Frequency *112*
  Generator Types *112*
Reactance *113*
  Inductive Reactance *113*
  Capacitive Reactance *115*
  Impedance *115*

Contents

> Ohm's Law for AC *116*
> Power Factor *116*
> Power versus Energy *117*
> > Power and Current *118*
>
> Single Phase *119*
> > Summary of Formulas for Single-Phase Circuits *119*
> > Voltage and Current Phase Relationships *120*
> > > Current, 121 Voltage, 122
> >
> > Real Power versus Reactive Power *124*
> > Summary of Formulas Involving Power *124*
>
> VA and KVA *125*
> Vars *126*
> > Single-Phase KVA *126*
> > Three-Phase KVA *126*
>
> Input Power (KVA) versus Output Power (HP) *128*
> Three Phase *129*
> > Power in Three-Phase Circuits *129*
> > Current in Three-Phase Circuits *130*
>
> Efficiency *131*
> > Current in Three-Phase Circuits *131*
> > Voltage in Three-Phase Circuits *132*
> > Power Delivered to a Three-Phase Motor *133*
> > Two-Phase Operation *133*
> > Two-Phase Motor Operating from a Three-Wire Line *133*
>
> Voltage Transformers: Step Up and Step Down *134*
> Current Transformers: Step Up and Step Down *134*

4 FORMULAS FOR MAGNETIC CIRCUITS *136*

> CGS, MKS, and English Systems *136*
> > The MKSA System and SI Units *137*
> > Standard Second *137*
> > The English, or Practical, System *138*
>
> English and Metric Units *138*
> Flux Lines *140*
> Ohm's Law for Magnetic Circuits *141*
> The Maxwell *143*
> Magnetic Force Around a Conductor *144*
> Magnetic Force Between Poles *145*
> Force on a Conductor *145*
> The DC Magnetic Field *146*
> The AC Magnetic Field *146*
> > Shape of Magnetic Lines *147*

Flux Density *147*
Reluctance *148*
   Reluctances in Series *149*
   Reluctances in Parallel *150*
Permeability *150*
   Relative Permeability *152*
Reluctivity versus Permeability *152*
Permeance *153*
Magnetomotive Force *154*
The Ampere-Turn *155*
Manipulating Magnetic Formulas *155*
Magnetic and Non-Magnetic Materials *157*
   Ferromagnetic *157*
   Paramagnetic *157*
   Diamagnetic *157*
Magnetic Aging *158*
   Core Loss Aging Coefficient *158*
   The Curie Point *158*
Magnetic Saturation *158*
Remanence *159*
Residual Induction *159*
Retentivity *159*
Coercive Force *159*
Hysteresis *160*
Hysteresis Loss *161*

5 DC MOTORS   *162*

Right-Hand Motor Rule *162*
The Commutator *163*
   Cleaning the Commutator *164*
Brushes *164*
   Brush Materials *164*
      Hard Carbon, *164* Electro-Graphitic, *164* Graphite, *167* Metal Graphite, *167*
   Brush Selection *167*
   Electrolytic Action *167*
   Brush Holders *167*
The Armature *168*
   Number of Turns per Armature Coil *169*
   The Armature's Magnetic Field *170*
Types of Windings *172*
   Lap Winding *173*
      Progressive Winding *175*

Contents   xi

Coil Pitch   *175*
Front Pitch and Back Pitch   *175*
Commutator Pitch   *175*
Retrogressive Winding   *176*
Simplex Winding   *176*
Duplex   *177*
Triplex   *177*
Simplex Progressive Lap Winding   *177*
Total Lap Winding   *178*
Symmetrical and Non-Symmetrical Connections   *179*
Wave Windings   *180*
Armature Poles   *180*
   Force Exerted on Armature Poles   *181*
   Eddy Current Power Loss in Laminated Cores   *183*
Field Coils   *183*
Interpoles   *184*
Motor Frames   *184*
   Open Frame   *184*
   Totally Enclosed Frame   *185*
   Protected Frame   *185*
   Drip-Proof Frame   *185*
   Splash-Proof Frame   *185*
   Dust-Proof Frame   *186*
   Watertight Frame   *186*
   Explosion-Proof Frame   *186*
   Frame Variations   *186*
      Heat Dissipation, 186   Vented, 186   Internal Fan Cooled, 186   Cooling Band, 186   Totally Enclosed, Fan Cooled, 186   Open Construction, 186   Motor Frame with Conduit Box, 186   Mounts, 187
Motor Bearings   *187*
   Bearing Oil   *187*
Motor Shaft   *187*
Series-Wound Motor   *188*
   Speed Control   *188*
Split-Field Series-Wound Motor   *190*
Shunt-Wound Motor   *190*
Field Current in a Shunt Motor   *192*
Motor Operating Voltage   *193*
Compound-Wound Motor   *194*
Differentially Wound Compound Motor   *195*
Cumulative-Wound Compound Motor   *196*
Transmission of Power   *198*
Name Plate (Data Plate)   *199*
Stepless Motors   *199*

Stepping Motors *199*
   Step versus Step Angle *200*
   Step Frequency *200*
   Microsteps *200*
   Incremental Mode *200*
   Types of Stepping Motors *199*
      Disk Stepper, 200   Variable Reluctance (VR), 201   Permanent-Magnet (PM) Rotor, 201   Hybrid, 203
   Terminology *204*
Insulation *204*
Motor Symbols *205*
Brushless DC Motors *205*
   Applications *207*
   Operational Setup *207*
   Slotless Stator *207*
   Miscellaneous Data *208*

6 AC MOTORS *209*

Basic AC Motors *209*
The Universal Motor *211*
   Speed Control *213*
      Compensated Series Wound *214*
      Distributed Field Compensated *214*
Plug and Socket Symbols *214*
Single-Phase and Polyphase Motors *214*
Repulsion Motors *215*
   Repulsion Motors *215*
   Repulsion Start Induction *215*
   Repulsion Induction *216*
The Squirrel-Cage Concept *217*
Single-Phase Motors *219*
Power-Line Losses *220*
   Power-Line Frequency *220*
      Current Requirements for Single-Phase Motors *222*
      Current versus Electrical Power Input *223*
Shaded-Pole Motor *224*
   Summary of the Characteristics of the Shaded-pole Motor *225*
Synchronous Motors *226*
   Self-Starting Induction-Reaction Synchronous Motor *226*
Split-Phase Motors *227*

Contents xiii

      Self-Starting Techniques for Split Phase   *228*
      Reversing Direction   *228*
      Starting versus Running Current   *228*
      Armature RPM   *229*
      Speed Variation   *230*
      Capacitor Split-Phase Motors   *230*
          Symbol, *232*
      Capacitor Start Two-Speed Motor   *232*
      Two-Capacitor Motor   *233*
      Permanent Capacitor Motor   *233*
      Autotransformer Effect   *234*
      Split-Capacitor Motor   *235*
Centrifugal Switches   *236*
Polyphase Motors   *238*
      Advantages of Polyphase   *238*
      Polyphase Motor Windings   *238*
      Polyphase Induction Motors   *240*
          Wound-Rotor Induction Motor, *240*
      Synchronous Speed   *240*
      Rotor Slip   *241*
      Current Requirements of Polyphase Motors   *242*
      Current in Three-Phase Circuits   *246*
      Horsepower, Efficiency, and Power Factor in Three-Phase Motors   *246*
Efficiency   *246*
Power   *248*
      Power Delivered to a Three-Phase Motor   *248*
      Two-Phase Motor Operating from a Three-Wire Line   *248*
      Terminal Connections for 120-V/240-V Motors   *249*

7 ELECTRIC/ELECTRONIC MOTOR CONTROL CIRCUITS   *250*

Electronic Power Supplies   *250*
      Autotransformer Speed Control   *251*
      Three-Phase Delta-Wye Power   *252*
Diac   *255*
Triac   *255*
Silicon-Controlled Rectifier (SCR)   *255*
SCR Speed Control   *255*
Chopper Action   *256*
Speed Control of Battery-Operated Series-Wound DC Motors   *256*
Speed Control for Universal Motor   *257*

Direction of Rotation of AC Motors   *257*
   Single-Phase Capacitor Motor   *258*
   Single-Phase Series Motor   *259*
   Split-Phase Motors   *259*
   Two-phase Motors   *259*
Fractional-Horsepower Motors   *260*
   Reversing the DC Shunt-Wound Motor   *261*
   Reversing the DC Compound-Wound Motor   *261*
   Reversing the Series-Wound Motor   *261*
   Reversing Polyphase Induction Motors   *261*
Speed and Regulation Control   *263*
   Motor Braking   *264*
   Diac and Triac Speed Control   *265*
Elimination of Transient Voltages   *266*
RFI Filter   *266*
   Induction Motor Control   *268*
   Motor-Reversing Circuitry   *269*
   Torque Control   *269*
   High-Current, Low-Speed Voltage Regulator   *270*
   Wheatstone Bridge Control   *270*
   Tachometer Motor Control   *271*
   High-Frequency Motor Control   *274*
   Series Motor Speed Control   *275*
   Sensor-Servo Control   *276*
Micromotors   *277*

8 SOLVED MOTOR PROBLEMS   *278*

INDEX   *295*

Preface

By its nature a data book is a reference book and is used to supply specific information about a selected topic or a formula for the solution of a problem. This work follows that same general procedure but, where required, supplies explanatory material. Still, certain assumptions are made.

One of these is that the reader has studied, or is in the process of becoming acquainted with, certain background material. This would include courses in basic algebra and, hopefully, trigonometry as well. This doesn't mean these subjects are absolutely essential, but a knowledge of these topics will be very helpful.

At one time the subject of motors was based only on an understanding of electricity, but now electronics has become an integral part. In some instances motors are actually made of electronic components and differ radically from those that are so well known. Electronics is also used as an adjunct to motors, often working as motor controls, but in either case a basic knowledge of both electricity as well as electronics will help make the use of this book both easier and more understandable.

The presentation pattern of subject material has been arranged so as to group related data. The earlier chapters can be considered as a review while the later chapters discuss DC, AC, and electronic motors.

Numerous examples are given for several reasons. They do help supply an understanding of the subject material but they are also useful as a guide for those preparing to become electrical/electronic technicians with motors as a specialty, for those preparing to become journeymen electricians, or to work in motor repair. Problems involving motors and their solutions are not only supplied throughout the book, but a separate chapter, with examples and solutions, is provided.

The first chapter supplies basic data insofar as that data relates to motors. While it may be considered a review it should serve as a fundamental background prior to a more detailed study of motors.

With few exceptions motors require a voltage input from a DC or AC source. There are various ways of supplying a source voltage, including batteries (detailed in Chapter 2), AC power lines, power supply rectifiers and filters, and motor generators. The fact that a motor is identified as DC, does not mean the AC power line is automatically excluded. The keynote of motors today are their high adaptability to practically any type of voltage source.

A motor is a link, working between its source voltage and a load. Since no motor is a solo performer it is essential to understand the various components that can supply an energy input, and since AC is one of the prime sources, it is discussed early, in Chapter 3.

Most students of electronics emphasize electrical, rather than magnetic, concepts. Chapter 4 explains basic magnetic theory and while it is fairly simple, it is essential for an understanding of motors of all types. This chapter should be regarded as a prerequisite for Chapter 5, DC motors, and Chapter 6, AC motors.

Controls for motors range from a simple on-off switch to elaborate arrangements, most often electronic, for detailed control. This involves turning the motor on, regulating armature speed, reversing the armature if necessary, braking the rotor, and making the armature step through preselected angles. While the circuit diagrams used for motor control are explained in Chapter 7 the ability to read such diagrams, and an understanding of solid-state electronics, will be helpful.

Martin Clifford
North Lauderdale, FL 33068

Acknowledgments

Though writing a technical book is a solitary process, it is accompanied by the need to rely on outside sources for data. Some manufacturers were reluctant to part with information, quite understandably, since it might have included privileged material. Fortunately, there were enough of those who could, and did, open their files. Not only manufacturers, but one group, the Small Motor Manufacturers Association, and one magazine, *Radio Electronics,* were highly cooperative.

I want to thank all of them and extend my gratitude for the help they supplied.

Cramer Co.
Eastern Air Devices
General Electric Co.
Penton/PC, Inc.
Portescap
RCA Corp.

Martin Clifford

chapter one

Basic Data

Every electrical motor is a transducer. In this case, the electrical motor transduces electrical energy to its mechanical equivalent. The electrical input can be supplied by one or more batteries, a power supply, an AC source, or by a DC generator. The output is in terms of a rotating shaft, which is expressed in horsepower (hp). There is always a relationship between the electrical input, the motor, and the turning force, or *torque,* of the shaft.

The operation of the motor can be described in terms of voltage, current, resistance, or electrical power, with these four interrelated.

VOLTAGE

Voltage is electrical pressure, also known as electromotive force (EMF), electrical pressure, potential gradient, voltage drop, IR drop, and potential difference.

Units of Voltage: The Volt

The basic unit of electrical pressure is the volt (V), but there are multiples, such as the kilovolt (kV) and megavolt (MV), and submultiples, such as the millivolt (mV) and microvolt (μV). There is a relationship between voltage and current, but a large voltage doesn't necessarily mean a large current.

VOLTAGE CONVERSIONS

Conversion from a larger to a smaller voltage involves multiplication; from a smaller to a larger quantity requires division. Table 1-1 summarizes voltage conversions.

Example:

Convert kilovolts (kV) to microvolts (μV).

Solution: Locate kV in left column. Move to right and locate 1,000,000,000 in column headed by microvolts: 1 kV = 1,000,000,000 μV.

Basic units such as the volt are used in formulas, making it necessary to work back and forth between submultiple units such as the microvolt and millivolt and the basic volt. It is sometimes also necessary to convert from millivolts to microvolts or vice versa.

A factor of 1,000, or 1,000,000, or 1,000,000,000 is used either as a multiplier or as a divisor. Division by 1,000 is done by moving the decimal point three places to the left, multiplication by 1,000 by moving the decimal point three places to the right. A similar procedure is followed with 1,000,000 and 1,000,000,000, except that for 1,000,000 the decimal point is moved six places, and for 1,000,000,000 it is moved nine places.

Example:

Divide 65,789 by 1,000.

Solution: Decimal points are often implied, rather than written: "65,789" is the same as "65,789." To divide by 1,000 move the decimal point three places to the left to obtain 65,789. To multiply by 1,000 move the decimal point three places to the right: 65,789 × 1,000 = 65,789. × 1,000 = 65,789,000.

For making number conversions it is often convenient to use exponents, as shown in Table 1-2.

TABLE 1-1. VOLTAGE CONVERSIONS

	mV	μV	kV	MV
mV	—	× 1,000	÷ 1,000,000	÷ 1,000,000,000
μV	÷ 1,000	—	÷ 1,000,000,000	÷ 1,000,000,000,000
kV	× 1,000,000	× 1,000,000,000	—	÷ 1,000
MV	× 1,000,000,000	× 1,000,000,000,000	× 1,000	—

V = volts; mV = millivolts; μV = microvolts; kV = kilovolts; MV = megavolts

Voltage Conversions

TABLE 1-2. VOLTAGE CONVERSIONS USING EXPONENTS

1 volt	$= 1,000$ mV $= 10^3$ mV $= 10^6$ μV
1 volt	$= 1,000,000$ μV $= 10^6$ μV
1 volt	$= 0.001$ kV $= 10^{-3}$ kV
1 volt	$= 0.000001$ MV $= 10^{-6}$ MV
1 kilovolt	$= 1,000$ V $= 10^3$ V
1 kilovolt	$= 0.001$ MV $= 10^{-3}$ MV
1 megavolt	$= 1,000,000$ V $= 10^6$ V
1 megavolt	$= 1,000$ kV $= 10^3$ kV
1 millivolt	$= 0.001$ V $= 10^{-3}$ V
1 millivolt	$= 1,000$ μV $= 10^3$ μV
1 microvolt	$= 0.000001$ V $= 10^{-6}$ V
1 microvolt	$= 0.001$ mV $= 10^{-3}$ mV

The Microvolt

The microvolt (μV), a submultiple of the basic volt, is a millionth of a volt, or 0.000001 V. In terms of exponents, it is 10^{-6} V. It requires 1,000,000 μV to equal 1 V.

$$1 \text{ V} = 1,000,000 \text{ μV} = 10^6 \text{ μV}$$

$$1 \text{ μV} = 1/1,000,000 \text{ V} = 0.000001 \text{ V} = 10^{-6} \text{ V}$$

The Millivolt

The millivolt (mV) is also a submultiple of the basic volt and is a thousandth of a volt, or 0.001 V. Using exponents it is 10^{-3} V. It requires 1,000 mV to equal 1 V.

$$1 \text{ V} = 1,000 \text{ mV} = 10^3 \text{ mV}$$

$$1 \text{ mV} = 1/1,000 \text{ V} = 0.001 \text{ V} = 10^{-3} \text{ V}$$

In terms of microvolts:

$$1 \text{ mV} = 1,000 \text{ μV} = 10^3 \text{ μV}$$

$$1 \text{ μV} = 1/1,000 \text{ mV} = 0.001 \text{ mV} = 10^{-3} \text{ mV}$$

The Kilovolt

The kilovolt (kV) is a multiple of the basic volt. The prefix *kilo* is a multiplier having a value of 1,000: 1 kV is equivalent to 1,000 V.

$$1 \text{ kV} = 1,000 \text{ V} = 10^3 \text{ V}$$

$$1 \text{ V} = 1/1,000 \text{ kV} = 0.001 \text{ kV} = 10^{-3} \text{ kV}$$

In terms of microvolts:

$$1 \text{ kV} = 1,000 \times 1,000,000 = 1,000,000,000 \text{ μV} = 10^3 \times 10^6 = 10^9 \text{ μV}$$

The Megavolt

A megavolt (MV) is the equivalent of 1,000,000 V, but is rarely used in connection with motors.

$$1 \text{ MV} = 1,000,000 \text{ V} = 10^6 \text{ V}$$

ABBREVIATIONS

The letter *V* is logically used as an abbreviation for volts, and so a designation of *200 volts* can be written as "200 V." Another letter *E* is also used to represent voltage, although it is actually an abbreviation for *electromotive force*. *E* is used to indicate the force itself, while *V* is used with a number to denote a specific quantity. Thus, $E = 200$ V.

Sometimes a letter is used as a multiplier. The lowercase letter *k* (kilo) indicates multiplication by 1,000. The value 1,000 V can be written as 1 kV. M (mega) represents 1×10^6, so 1×10^6 V can be expressed as 1 MV. A lowercase Greek letter μ ("mu") is an abbreviation of the prefix *micro;* it means "one-millionth," or 1×10^{-6}. A microvolt, written as "μV," is a millionth of a volt.

A number of electrical and electronic units of measure are capitalized, while others are not. Capital letters are used in basic units to honor those who have made important contributions to the study of electricity or the field of electronics. Thus, mA, dB, mW, kV, and mH are used as abbreviations for milliampere (after André-Marie Ampère), decibel (after Alexander Graham Bell), milliwatt (after James Watt), kilovolt (after Alessandro Volta), and millihenry (after Joseph Henry).

Example:

The voltage drop along a wire leading from a power source to a motor is 0.018 V. This is equivalent to how many millivolts?

Solution: To change volts to millivolts, multiply volts by 1,000, easily done by moving the decimal point three places to the right. 0.018 V = 18 mV.

SOURCE VOLTAGE

A source voltage is any voltage used to operate a motor or electronic circuit triggering a motor, or both, and can be direct current (DC) or alternating current (AC).

DC Voltage Symbols

The symbol for a battery is commonly used to represent a DC voltage, even though that voltage may be furnished by an electronic power supply. Battery symbols consist of pairs of parallel lines, with one line longer than the other. The longer

Current

of the two lines is "plus" or positive; the shorter "minus" or negative. Polarity designations such as + and − may or may not be included. In some instances just the plus sign is used. Voltage symbols are not standardized, but those in Figure 1-1 are typical.

CURRENT

The basic unit of electrical current is the ampere. Motor current is expressed in either the basic unit or in submultiples such as the milliampere and the microampere. Multiples larger than the ampere are not used.

The letter A is used to indicate currents of one ampere or more. Thus a motor operating on a two-ampere current may have a data plate reading "2 A."

The Milliampere

The milliampere (mA) is one-thousandth of an ampere. To convert milliamperes to amperes, divide milliamperes by 1,000, or move the decimal point three places to the left. Conversely, to convert amperes to milliamperes, multiply amperes by 1,000 or 10^3.

$$1 \text{ mA} = 0.001 \text{ A} = 1/1,000 \text{ A} = 10^{-3} \text{ A}$$

$$1 \text{ A} = 1,000 \text{ mA} = 10^3 \text{ mA}$$

Example:

What is the current rating of a motor in milliamperes if it uses an operating current of 2.87 A?

Solution:

$$2.87 \text{ A} = 2.87 \times 1,000 = 2,870 \text{ mA, or}$$

$$2.870 = 2,870. \text{ mA (moving the decimal point three places to the right).}$$

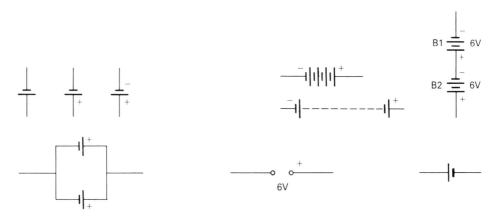

Figure 1-1. Battery or DC voltage symbols.

The Microampere

There are some motors and associated circuits that require currents less than 1 mA. While the milliampere as a current unit could be used to describe such currents, it would involve unhandy fractions such as 3/17 mA or 28/93 mA. This could be simplified by using decimals, but would then involve awkward numbers requiring many zeros. It could result in currents such as 0.000032 mA or 0.0053 mA.

To be able to use more convenient whole numbers, another submultiple of the ampere is available and is known as the microampere, represented by μA. The value 0.01 mA, for example, could be written as "10 μA."

The microampere is equivalent to one millionth of an ampere or a thousandth of a milliampere. Divide an ampere into a million equal currents and each would be 1 μA. Or, divide a current of 1 mA into a thousand equal currents and each would be 1 μA.

Current Conversion Rules

To convert amperes to milliamperes, multiply amperes by 1,000. To convert amperes to microamperes, multiply amperes by 1,000,000. To move in the other direction, divide. To convert milliamperes to amperes, divide milliamperes by 1,000. To convert microamperes to amperes, divide microamperes by 1,000,000. These steps are summarized in accompanying Tables 1-3 and 1-4.

TABLE 1-3. CURRENT CONVERSIONS

	A	mA	μA
A	—	× 1,000 or × 10^3	× 1,000,000 or × 10^6
mA	÷ 1,000 or × 10^{-3}	—	× 1,000 or × 10^3
μA	÷ 1,000,000 or × 10^{-6}	÷ 1,000 or × 10^{-3}	—

A = amperes; mA = milliamperes; μA = microamperes

TABLE 1-4. ALTERNATIVE ARRANGEMENT OF CURRENT CONVERSIONS

1 ampere = 1,000 mA = 10^3 mA = 10^6 μA
1 ampere = 1,000,000 μA = 10^6 μA
1 milliampere = 1/1,000 A = 10^{-3} A
1 milliampere = 1,000 μA = 10^3 μA = 1/1,000 A = 10^{-3} A
1 microampere = 1/1,000,000 A = 10^{-6} A
1 microampere = 1/1,000 mA = 10^{-3} mA

Current

Example:

What is the equivalent current in microamperes of a current that is measured at 22 mA?

Solution:

$$22 \text{ mA} \times 1{,}000 = 22{,}000 \text{ microamperes} = 22{,}000 \text{ μA}$$

$$22 \text{ mA} = 22. \text{ mA}$$

Move the decimal point three places to the right: 22.0 mA = 22,000 microamperes = 22,000 μA.

Example:

In a subminiature motor a current of 3,645 μA flows in the armature. What is the equivalent current in milliamperes and in amperes?

Solution: To convert to milliamperes move the decimal point three places to the left.

$$3{,}645. \text{ μA} = 3.645 \text{ mA}$$

To convert to amperes, divide by one million or move the decimal point six places to the left.

$$3{,}645. \text{ μA} = 0.003645 \text{ A, or}$$

$$3{,}645. \text{ μA} = 3{,}645/1{,}000{,}000 \text{ A}$$

Current Designations

The phrase *electrical current* is general and is applicable overall. Sometimes more specific information is needed, and currents are thus often identified by function. The following list is indicative of some current types.

Line current	Sinusoidal current
Field current	Non-sinusoidal current
Armature current	Starting current
Eddy current	Operating current
Direct current	Overload current
Alternating current	Maximum current
Pulsating direct current	Minimum current
Pulsating alternating current	Bypass current
Single-phase current	Base current
Polyphase current	Emitter current
Leading current	Collector current
Lagging current	Sawtooth current

RESISTANCE

Resistance, expressed in a basic unit called the *ohm* (Ω), is the opposition to the flow of a current in a resistive circuit. While there are certain components specifically designed to provide resistance, a coil of wire such as an armature, or the connecting leads from a motor to a power source, also have resistance.

Although there are submultiples of resistance, multiples such as the kilohm and the megohm are more commonly used.

The Kilohm

The kilohm (kΩ) is equal to a thousand ohms. Thus:

$$1 \text{ k}\Omega = 1,000 \text{ }\Omega = 10^3 \text{ }\Omega$$

$$1 \text{ }\Omega = 1/1000 \text{ k}\Omega = 0.001 \text{ k}\Omega = 10^{-3} \text{ k}\Omega$$

The Megohm

The megohm (MΩ) is another multiple of the ohm and is equal to a million ohms.

$$1 \text{ M}\Omega = 1,000,000 \text{ }\Omega = 10^6 \text{ }\Omega$$

$$1 \text{ }\Omega = 1/1,000,000 \text{ M}\Omega = 0.000001 \text{ M}\Omega = 10^{-6} \text{ M}\Omega$$

It is sometimes necessary to work between the kilohm and the megohm.

$$1 \text{ k}\Omega = 1/1,000 \text{ M}\Omega = 0.001 \text{ M}\Omega = 10^{-3} \text{ M}\Omega$$

$$1 \text{ M}\Omega = 1,000 \text{ k}\Omega = 10^3 \text{ k}\Omega$$

Table 1-5 is a summary of resistance conversions. The letter *R* is used to identify resistance.

The same methods used for making voltage and current conversions are also applicable to resistance. Table 1-6 lists the basic and multiple values of resistance.

The amount of resistance covers an extremely wide range and extends from the abohm, (nanohm) 10^{-9} ohm or a billionth of an ohm, to a gigohm, 10^9 ohms or a billion ohms. The designation *begohm,* for a billion ohms, is no longer used.

TABLE 1-5. RESISTANCE CONVERSIONS

	Ω	kΩ	MΩ
Ω	—	$\div$ 1,000 or $\times$ 10^{-3}	$\div$ 1,000,000 or $\times$ 10^{-6}
kΩ	$\times$ 1,000 or $\times$ 10^3	—	$\div$ 1,000 or $\times$ 10^{-3}
MΩ	$\times$ 1,000,000 or $\times$ 10^6	$\times$ 1,000 or $\times$ 10^3	—

Ω = ohms; kΩ = kilohms; MΩ = megohms.

Resistance

TABLE 1-6. BASIC AND MULTIPLE VALUES OF RESISTANCE

1 ohm	$= 10^{-3}$ kΩ = 0.001 kΩ = 1/1,000 kΩ
1 ohm	$= 10^{-6}$ MΩ = 0.000001 MΩ = 1/1,000,000 MΩ
1 kilohm	$= 10^{3}$ Ω = 1,000 Ω
1 kilohm	$= 10^{-3}$ MΩ = 0.001 MΩ = 1/1,000 MΩ
1 megohm	$= 10^{3}$ kΩ = 1,000 kΩ
1 megohm	$= 10^{6}$ Ω = 1,000,000 Ω

These are extremes and the resistances in motors and motor control electronic circuits are well with these limits. For motors and motor circuits resistance values are from 1 Ω (but usually more) to 10 MΩ (but usually less).

An abohm and a gigohm are beyond the capabilities of most resistance testing instruments, so from a practical point of view an abohm is a short circuit, a gigohm an open circuit.

Resistor Power Designations

The wattage rating of a resistor is an indication of its ability to radiate heat; resistors are designated from 1/2 watt (or less) to 100 watts (or more). The high-power resistors may be found in starters for DC motors; the low-power units in electronic control circuits for both DC and AC motor types.

Resistor Types

Resistors can be identified not only by their resistance rating but also by their resistance tolerance, that is, the amount of resistance deviation from that specified. The deviation is specified in terms of a percentage. Resistors are also known by their usage, shape, physical construction, and whether they are fixed, tapped, or variable. Table 1-7 indicates types of resistors.

Table 1-8 lists some of the more commonly used resistors, indicating some of their physical and electrical characteristics.

TABLE 1-7. RESISTOR TYPES

Radial	Trimmer
Axial	Cermet
Metal Oxide	Miniature
Non-Inductive	PC Board
Carbon Composition	Wirewound
Carbon Film	Potentiometer
Metal Film	Power
Deposited Film	Precision
Ceramic	Conductive Plastic
Chip	Hybrid
Fixed	Surface Mount
Variable	

TABLE 1-8. CHARACTERISTICS OF RESISTORS

Carbon composition	**Resistance range:** 2.7 ohms to 100 megohms **Power rating:** to 2 watts **Tolerance:** 1% to 20% **Temperature coefficient:** −200 to −8,000 PPM/°C **Notes:** General purpose. Excellent transient and surge handling capabilities. Resistance increases by 20% during storage under humid conditions. Made from a mixture of graphite and a nonconductive binder such as clay to hold the particles in place. Resistance determined by the ratio of carbon to binder material, with the resistance varying from less than 1 ohm to several megohms.
Carbon composition potentiometer	**Resistance range:** 50 ohms to 10 megohms **Power rating:** to 5 watts **Temperature coefficient:** 1,000 PPM/°C **Life expectancy:** 5,000,000 rotations **Notes:** High shaft torque causes poor adjustability. The resistive element is carbon composition or metallic film for potentiometers and resistance wire for rheostats. In some instances a conductive ceramic element is used for potentiometers. **Concentric Potentiometer** Dual variable resistor equipped with two shafts, an inner and an outer. Each can be operated independently. **Fixed-Film Resistors** Made of a core of a nonconductive material on which is sprayed a thin layer or film of a resistive substance. They have a higher power rating than carbon-composition resistors and are sometimes used as a substitute for wirewound types. The film can be applied in a continuous layer with the resistance determined by the thickness of the film. This film can be made of carbon whose particles are micro-crystalline, and is sometimes mixed with boron or various metal oxides.
Carbon film	**Resistance range:** 10 ohms to 25 megohms **Power rating:** 0.1 to 10 watts **Tolerance:** 2% to 10% **Temperature coefficient:** −200 to −1,000 PPM/°C **Notes:** General purpose, cost less than carbon-composition units.
Metal film	**Resistance range:** 10 ohms to 3 megohms (high voltage types: 1 kilohm to 30 gigohms) **Power rating:** to 10 watts (high voltage types: to 6 watts) **Tolerance:** 0.1% to 2% **Temperature coefficient:** ±25 to ±175 PPM/°C **Life expectancy (potentiometers):** 100,000 rotations **Failure mode:** resistance change or catastrophic failure **Notes:** Fair degree of precision in lower value units. High stability, long life.
Film networks	**Resistance range:** 10 ohms to 33 megohms **Power rating:** to 0.2 watts per element, to 1.6 watts per network **Tolerance:** 0.1% to 5% **Operating temperature range:** −55 to +125°C **Temperature coefficient:** ±25 to ±300 PPM/°C
Chip resistors	**Resistance range:** 1 ohm to 100 megohms **Power rating:** to 2 watts **Tolerance:** 1% to 20% **Operating temperature range:** −55 to +125°C

Resistance

Power wirewound
: **Resistance range:** 0.1 ohm to 180 kilohms
Power rating: to greater than 225 watts
Tolerance: 5% to 10%
Temperature coefficient: less than ±260 PPM/°C
Notes: Wirewound resistors consist of resistance wire wound around a nonconductive core made of a heat resistive material such as ceramic, with the resistance dependent on the type of wire used. Their advantage is their higher power dissipation capability over resistors such as the carbon type.

Precision wirewound
: **Resistance range:** 0.1 ohm to 800 kilohms
Power rating: to 15 watts
Tolerance: .01% to 1%
Life expectancy (potentiometers): 2000,000 to 1,000,000 rotations
Notes: Used in low-tolerance, high-power dissipation applications where AC performance is not critical. Power dissipation depends on heat-sink or air flow around the device. When mounting on a PC board, use standoffs to prevent charring the board. Wirewound potentiometers do not suffer from contact resistance variations. The units can be manufactured with low temperature coefficients and tight tolerances. Applications include motor speed controls. Precision types used in servo mechanisms.

Cermet
: **Resistance range:** 50 ohms to 5 megohms
Power rating: to 2 watts
Life expectancy (potentiometers): 50 to 500,000 rotations
Notes: Very stable under humid conditions. Low temperature coefficients. Low end resistance (2 ohms). Short life expectancy. Cermet is also the thick film used in resistor networks and chip resistors.

Ceramic
: **Life expectancy:** Long in-service life.
Notes: Made of carborundum, a carbon-silicon compound used as a diode demodulator in the early days of radio. Affected by the voltage across it with its resistance varying inversely with the applied EMF. Made having either a positive or negative temperature coefficient. Available in fixed or variable form. Also known as a voltage-dependent resistor.

Conductive plastic potentiometers
: **Resistance range:** 150 ohms to 5 megohms
Power rating: to 1 watt
Temperature coefficient: −600 to −300 PPM/°C
Life expectancy: 100,000 to 4,000,000 rotations

General purpose conductive plastic potentiometers
: **Resistance range:** 1 ohm to 15 kilohms, depending on power rating
Power rating: to 1,000 watts

Precision conductive plastic potentiometer
: **Resistance range:** 100 ohms to 500 kilohms
Power rating: to 7 watts
Tolerance: 3%
Life expectancy: Greater than 2,000,000 rotations

Conductive plastic trimmers
: **Resistance range:** 10 ohms to 100,000 ohms
Power rating: to 1 watt
Notes: Plastic potentiometers have a long life expectancy. Resistance will shift if exposed to humidity.

Hybrid potentiometers
: **Resistance range:** 200 to 250,000 ohms
Power rating: to 7 watts
Tolerance: 5%
Life expectancy: 10,000,000 rotations

Codes and Values

Resistors are commonly used in control circuits associated with motors. To be able to identify and locate them they are sometimes assigned codes and values. R1 is resistor number 1; R2 is resistor number 2. An alphanumeric designation, such as R12, is called a *code;* the amount of resistance is referred to as the *value.* The abbreviation for ohm is the Greek uppercase letter omega (Ω). Codes and values can also be assigned to voltages and currents. Thus, E1, E2, and I1 and I2, and so on.

Table 1-9 shows the conversion relationship between the basic units of the volt, ohm, and ampere, their symbols, multiples, and values. A voltage can be written as E1 = 6 volts, or E1 = 6 V, and a current can be described as I3 = 2 amperes or I3 = 2 A.

Resistor symbols. Resistors can be fixed, tapped, or variable. Symbols for these are shown in Figure 1-2.

Resistor color code. Fixed resistors can have their ohmic value identified by colors encircling the resistor at one end. The first three colors, Figure 1-3, represent the total resistance. The fourth color, if supplied, is the tolerance. Table 1-10 lists the colors and their values.

TABLE 1-9. SUMMARY OF *E, I,* AND *R* CONVERSIONS

Unit	Symbol	Multiple	Value
		In Terms of Numbers	
volt	E	kilovolt (kV)	1,000 volts
volt	E	millivolt (mV)	1/1,000 volt
volt	E	microvolt (μV)	1/1,000,000 volt
ohm	R	kilohm (kΩ)	1,000 ohms
ohm	R	megohm (MΩ)	1,000,000 ohms
ampere	I	milliampere (mA)	1/1,000 ampere
ampere	I	microampere (μa)	1/1,000,000 ampere
		In Terms of Exponents	
1 volt		= 10^3 millivolts	= 10^6 microvolts
1 millivolt		= 10^{-3} volt	= 10^3 microvolts
1 microvolt		= 10^{-6} volt	= 10^{-3} millivolt
1 ohm		= 10^{-3} kilohm	= 10^{-6} megohm
1 kilohm		= 10^3 ohms	= 10^{-3} megohm
1 megohm		= 10^6 ohms	= 10^3 kilohms
1 ampere		= 10^3 milliamperes	= 10^6 microamperes
1 milliampere		= 10^{-3} ampere	= 10^3 microamperes
1 microampere		= 10^{-6} ampere	= 10^{-3} milliamperes

Resistance

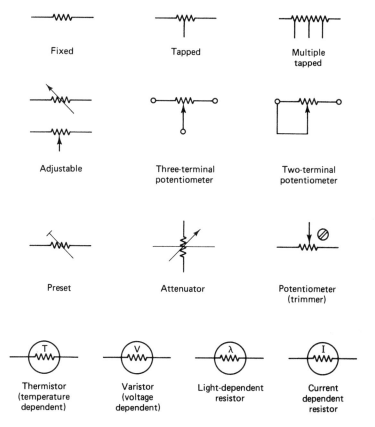

Figure 1-2. Fixed and variable resistors.

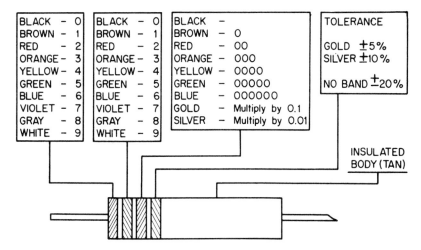

Figure 1-3. Color code for fixed carbon resistors. Hold the resistor so colors are at left.

TABLE 1-10. RESISTOR COLOR CODE

Color	First digit	Second digit	Multiplier	Tolerance ($\pm$)
Black	—	0	1	—
Brown	1	1	10	1%
Red	2	2	100	2%
Orange	3	3	1,000	3%
Yellow	4	4	10,000	4%
Green	5	5	100,000	
Blue	6	6	1,000,000	
Violet	7	7	10,000,000	
Gray	8	8	100,000,000	
White	9	9	1,000,000,000	
Silver				10%
Gold				5%
No color				20%

Example:

What is the color coding for a 5,600-Ω resistor?

Solution: A 5,600-Ω resistor is green (first color = 5); blue (second color = 6), and red (third color = 00). Since no fourth color was indicated, this resistor has a tolerance of $\pm 20\%$.

Tolerance

The resistor color code is just an approximation. A plus tolerance is indicated by plus (+), a minus by (−).

Example:

A 5,100-Ω resistor has a tolerance of $\pm 20\%$. What is the possible resistance range of this component?

Solution:

$$20\% \text{ of } 5,100 = 0.20 \times 5,100 = 1,020$$

$$5,100 + 1,020 = 6,120 = \text{upper resistance limit}$$

$$5,100 - 1,020 = 4,080 = \text{lower resistance limit}$$

This resistor can be anywhere in the range of 4,080 to 6,120 Ω.

Resistor Connections

Resistors can be connected in series, in parallel (shunt), or in series-parallel.

Resistance

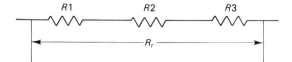

Figure 1-4. Resistors in series.

Resistors in Series. For resistors wired in series (Figure 1-4), the fact that they are adjacent doesn't mean they will be physically close when wired. The resistors may be identified by codes or by codes and values. The formula for this type of connection is:

$$R_t = R1 + R2 + R3 \ldots$$

where R_t is the total resistance and $R1$, $R2$, $R3$, and so on are the resistors. The resistors may be in ohms, kilohms, or megohms, but it is helpful to convert them to basic units (ohms) or identical multiples before problem solving. The resistors can be interchanged without affecting the total resistance.

Example:

Three resistors in series are 450 Ω, 1.2 kΩ, and 0.01 MΩ. What is the total resistance?

Solution:

$$1.2 \text{ k}\Omega = 1.2 \times 1000 = 1{,}200 \text{ }\Omega$$

$$0.01 \text{ M}\Omega = 0.01 \times 1{,}000{,}000 = 10{,}000 \text{ }\Omega$$

$$R_t = 450 + 1200 + 10{,}000 = 11{,}650 \text{ }\Omega$$

Two Resistors in Parallel. If the two resistors have identical values, the total resistance is half that of either one. That is:

$$R_t = \frac{R1}{2} \quad \text{or} \quad \frac{R2}{2}$$

If the resistors have different values, the total resistance can be calculated by either of these two formulas:

$$R_t = \frac{R1 \times R2}{R1 + R2}$$

$$\frac{1}{R_t} = \frac{1}{R1} + \frac{1}{R2}$$

Example:

A pair of resistors (Figure 1-5), 40 and 80 Ω, are wired in parallel. What is the equivalent resistance?

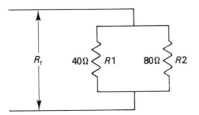

Figure 1-5. Two resistors in parallel.

Solution:

$$R_t = \frac{R1 \times R2}{R1 + R2}$$

$$= \frac{40 \times 80}{40 + 80} = \frac{3200}{120} = 26.67 \; \Omega$$

Alternative Solution:

$$\frac{1}{R_t} = \frac{1}{R1} + \frac{1}{R2}$$

$$= \frac{1}{40} + \frac{1}{80}$$

$$= 0.025 + 0.0125 = 0.0375$$

$$= \frac{1}{0.0375} = 26.67 \; \Omega$$

The 10 to 1 Rule. When two resistors are in parallel and one has 10 times (or more) the resistance of the other, the total resistance can be considered as the value of the smaller.

Example:

What is the equivalent value of two parallel resistors if one is 20 Ω and the other is 200 Ω?

Solution:

$$R_t = \frac{R1 \times R2}{R1 + R2} = \frac{20 \times 200}{20 + 200} = \frac{4000}{220} = 18.2 \; \Omega$$

Three or More Resistors in Parallel. There are several possible methods for combining three or more resistors in parallel. One technique is to use the formula for two parallel resistors, selecting any two of the three resistors, then combining the answer with the remaining resistor.

Figure 1-6 shows the circuit arrangement for three parallel resistors. There are various formulas available, any one of which can be used for determining the equivalent resistance.

Resistance

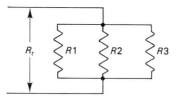

Figure 1-6. Three resistors in parallel.

For three (or more) resistors:

$$\frac{1}{R_t} = \frac{1}{R1} + \frac{1}{R2} + \frac{1}{R3} \ldots$$

For three (or more) resistors:

$$R_t = \frac{1}{\frac{1}{R1} + \frac{1}{R2} + \frac{1}{R3}} \ldots$$

or

$$R_t = \frac{R1R2R3}{R2R3 + R1R3 + R1R2}$$

For four (or more) resistors:

$$R_t = \frac{R1}{1 + R1/R2 + R1/R3 + R1/R4} \ldots$$

The resultant value of parallel resistors when only two are involved can be determined in three different ways: by the use of a formula, or by a table, such as Table 1-11 and 1-12, or by a nomogram.

Table 1-11 supplies the total resistance, R_t, for two resistors R1 and R2 in parallel, with resistance values from 1 to 68 V, while Table 1-12 covers the range from 10 to 680 Ω. For both tables, R1 and R2 must be in the same units: ohms, kilohms, or megohms.

To use either of these tables extend a line downward from the selected value of R1. Then extend a line horizontally from the value of R2. The point of intersection of these two lines is the desired value of R_t.

Resistors in Series-Parallel. The total resistance can be calculated by combining the resistors using the formulas for series and parallel combinations.

$$R_t = R1 + \frac{1}{\frac{1}{R2} + \frac{1}{R3} + \frac{1}{R4}}$$

TABLE 1-11. PARALLEL RESISTANCES FROM 1 TO 68 Ω

R1	1	1.5	2.2	3.3	4.7	6.8	10	15	22	33	47	68
R2												
1	0.50	0.60	0.69	0.77	0.83	0.87	0.91	0.93	0.95	0.97	0.98	0.99
1.5	0.60	0.75	0.89	1.03	1.14	1.22	1.30	1.36	1.40	1.43	1.45	1.46
2.2	0.69	0.89	1.10	1.32	1.50	1.66	1.82	1.92	2.00	2.06	2.10	2.13
3.3	0.77	1.03	1.32	1.65	1.94	2.22	2.48	2.70	2.87	3.00	3.08	3.14
4.7	0.83	1.14	1.50	1.94	2.35	2.78	3.20	3.58	3.87	4.12	4.27	4.39
6.8	0.87	1.22	1.66	2.22	2.78	3.40	4.05	4.58	5.79	5.64	5.94	6.18
10	0.91	1.30	1.80	2.48	3.20	4.05	5.0	6.0	6.9	7.7	8.3	8.7
15	0.93	1.35	1.92	2.70	3.58	4.68	6.0	7.50	8.9	10.3	11.4	12.2
22	0.95	1.40	2.00	2.87	3.87	5.19	6.9	8.90	11.0	13.2	15.0	16.6
33	0.97	1.43	2.06	3.00	4.12	5.64	7.7	10.3	13.2	16.5	19.4	22.2
47	0.98	1.45	2.10	3.08	4.27	5.94	8.3	11.4	15.0	19.4	23.5	27.8
68	0.99	1.47	2.13	3.14	4.39	6.18	8.7	12.2	16.6	22.2	27.8	34.0

TABLE 1-12. PARALLEL RESISTANCES FROM 10 TO 680 Ω

R1\R2	10	15	22	33	47	68	100	150	220	330	470	680
10	5.0	6.0	6.9	7.7	8.3	8.7	9.1	9.3	9.5	9.7	9.8	9.9
15	6.0	7.5	8.9	10.3	11.4	12.2	13.0	13.6	14.0	14.3	14.5	14.6
22	6.9	8.9	11.0	13.2	15.0	16.6	18.2	19.2	20.0	20.6	21.0	21.3
33	7.7	10.3	13.2	16.5	19.4	22.2	24.8	27.0	28.7	30.0	30.8	31.4
47	8.3	11.4	15.0	19.4	23.5	27.8	32.0	35.8	38.7	41.2	42.7	43.9
68	8.7	12.2	16.6	22.2	27.8	34.0	40.5	46.8	51.9	56.4	59.4	61.8
100	9.1	13.0	18.2	24.8	32.0	40.5	50	60	69	77	83	87
150	9.3	13.6	19.2	27.0	35.8	46.8	60	75	89	103	114	122
220	9.5	14.0	20	28.7	38.7	51.9	69	89	110	132	150	166
330	9.7	14.3	20.6	30.0	41.2	56.4	77	103	132	165	194	222
470	9.8	14.5	21.0	30.8	42.7	59.4	83	114	150	194	235	278
680	9.9	14.6	21.3	31.4	43.9	61.8	87	122	166	222	278	340

or for multiple series-parallel combinations

$$R_t = R1 + \frac{R2 \times R3}{R2 + R3} + R4 + R5 + \frac{R6 \times R7}{R6 + R7}$$

Figure 1-7 shows a single series resistor and three resistors in parallel. Figure 1-8 illustrates three series resistors and two pairs of parallel resistors.

Summary of Resistor Connections.

Resistors in Series

$$R_t = R1 + R2 + R3$$

Two Resistors in Parallel

$$R_t = \frac{R1 \times R2}{R1 + R2}$$

$$\frac{1}{R_t} = \frac{1}{R1} + \frac{1}{R2}$$

$$R_t = \frac{R1}{2} \text{ or } \frac{R2}{2} \text{ (when both resistors are identical)}$$

Three or More Resistors in Parallel

$$\frac{1}{R_t} = \frac{1}{R1} + \frac{1}{R2} + \frac{1}{R3} \ldots$$

$$R_t = \frac{1}{\frac{1}{R1} + \frac{1}{R2} + \frac{1}{R3}} \ldots$$

$$R_t = \frac{R1}{1 + R1/R2 + R1/R3} \ldots$$

Four or More Resistors in Parallel

$$R_t = \frac{R1}{1 + R1/R2 + R1/R3 + R1/R4} \ldots$$

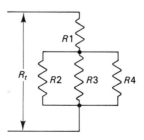

Figure 1-7. Single resistor in series with three parallel resistors.

Conductivity

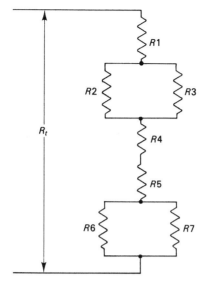

Figure 1-8. Series-parallel network.

Resistors in Series-Parallel

$$R_t = R1 + \cfrac{1}{\cfrac{1}{R2} + \cfrac{1}{R3} + \cfrac{1}{R4}}$$

$$R_t = R1 + \frac{R2 \times R3}{R2 + R3} + R4 + R5 + \frac{R6 \times R7}{R6 + R7}$$

RESISTIVITY OF METALS

The resistivity of a metal is its opposition to the flow of an electrical current in a cube of the substance measuring one centimeter on each side, measured at a specific temperature. Copper is one of the best conductors, both on a conductivity and a cost basis. Silver has slightly better conductivity, but is more expensive. Table 1-13 supplies the resistivity of metals in comparison with that of copper.

CONDUCTIVITY

Voltage and current are two quantities found in connection with motors. Currents are able to pass easily through some substances but only with great difficulty through others. A material that allows the comparatively easy passage of an electrical current through it is a conductor. The word *comparative* is used, since no two substances are equally conductive. Silver is an excellent conductor. Copper is almost as good. Aluminum and zinc are next in order. Motor brushes have good conductivity.

TABLE 1-13. RELATIVE RESISTIVITY OF METALS

Materials	Resistivity compared to copper
Aluminum (pure)	1.6
Brass	3.7-4.9
Cadmium	4.4
Chromium	1.8
Copper (hard-drawn)	1.03
Copper (annealed)	1.00
Gold	1.4
Iron (pure)	5.68
Lead	12.8
Nickel	5.1
Phosphor bronze	2.8-5.4
Silver	0.94
Steel	7.6-12.7
Tin	6.7
Zinc	3.4

Offhand it might seem desirable to have maximum conductivity, and under certain circumstances this is correct. Sometimes, however, it is just as desirable to reduce conductivity deliberately. The basic unit of conductivity is the *siemens* (symbol: S), formerly called the *mho*. Conductivity, identified by the letter G, is the reciprocal of resistance. Thus,

$$G = 1/R$$

Here G is the conductivity in siemens and R is the resistance in ohms.
The formula can be changed to find the value of R when G is known.

$$R = 1/G$$

For the most part resistance rather than conductivity is used when working with motors.

Summary of Conductances

Conductances in Parallel

$$G_t = G_1 + G_2 + G_3 \ldots$$

Ohm's Law for Conductances

$$E = I/G$$

$$I = E/G$$

$$G = \frac{1}{R}$$

$$R = \frac{1}{G}$$

INSULATORS

An insulator is a substance that has a very high opposition to the passage of an electrical current. Materials having insulating properties can be depended upon, more or less, to oppose the flow of an electric current through them. Mica is used to separate the copper bars of the commutator of a DC motor because of its excellent insulating properties.

VOLTAGE, CURRENT, AND RESISTANCE RELATIONSHIPS

The relationship between voltage, current and resistance in DC circuits can be expressed as

$$\text{voltage} = \text{current} \times \text{resistance}$$

Known as Ohm's Law, it can be simplified by using codes in place of words. The statement would then appear as

$$E = I \times R$$

where

E is the voltage in volts,

I is the current in amperes, and

R is the resistance in ohms.

The expression $E = I \times R$ can be rearranged:

$$I = E/R \text{ and } R = E/I$$

If the values of any of two terms are supplied, the third can be calculated. In these formulas, E is in volts, R is in ohms, and I is in amperes. If the data supplied are in milliamperes or microamperes, in microvolts, millivolts, or kilovolts, or in kilohms or megohms, these must first be converted to the basic units of amperes, volts, and ohms. However, no conversion is required if the data are all in equivalent units.

Example:

What is the current in a DC circuit if the EMF is 10 V and the resistance is 5 Ω?

Solution:

$$I = E/R = 10/5 = 2 \text{ A}$$

Since the terms are both in basic units—volts and ohms—the answer is also in a basic unit—amperes.

Example:

What is the resistance in a motor control circuit if the measured EMF is 10 V and the current is 1,200 mA?

Solution:

$$R = E/I = 10/1.2 = 8.33 \text{ }\Omega$$

Note that the current in milliamperes was first changed to its basic unit in amperes: $1,200/1,000 = 1.2$ A.

Example:

What is the voltage loss along a wire carrying a current of 120 mA and having a resistance of 0.08 Ω?

Solution:

$$E = I \times R = 0.120 \times 0.08 = 0.0096 \text{ V}$$

In this example the current of 120 mA was first converted to amperes by dividing it by 1,000 or by moving the decimal point three places to the left.

The triangle in Figure 1-9 is a memory aid for remembering Ohm's Law. Cover the unknown value and the formula is revealed. If the wanted data is voltage, cover it and the answer is shown below, I times R. Cover I and the answer is E over R. Similarly, if R is the unknown value, cover it and the formula is revealed as E over I.

Summary of Ohm's Law for DC

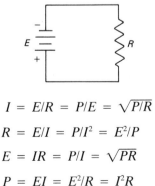

$$I = E/R = P/E = \sqrt{P/R}$$
$$R = E/I = P/I^2 = E^2/P$$
$$E = IR = P/I = \sqrt{PR}$$
$$P = EI = E^2/R = I^2R$$

IR Drops

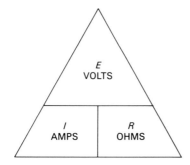

Figure 1-9. Memory device for Ohm's Law.

where P is the power in watts, E is the voltage in volts, I is the current in amperes, and R is the resistance in ohms.

IR DROPS

Motor circuits sometimes use resistors to obtain a desired amount of voltage when the DC source voltage is higher than required. In some instances, when there is more than one circuit, resistors can work as voltage dividers to supply each circuit with its specific required voltage. This could be done by using different voltage sources, but using resistors is more convenient and practical. Figure 1-10 shows two resistors, identified as R1 and R2, connected in series across a 24-V DC supply. The purpose here is to obtain 4.8 V for one motor circuit, 19.2 V for another.

In this diagram R1 is 12 Ω and R2 is 48 Ω. Their total is 60 Ω, so the current flowing in the circuit is 0.4 A. The voltage across each resistor, sometimes called an IR drop, is $E = I \times R = 0.4 \times 12 = 4.8$ V for R1, and $0.4 \times 48 = 19.2$ V for R2. The IR drops across R1 and R2 can be changed by using different values of resistance.

Polarity of IR Drops

When a current flows through a resistor the voltage or IR drop across that resistor has polarity, just as any other type of voltage. If the direction of current flow is known, the polarity can be marked on the resistor symbol, as in Figure 1-11. The

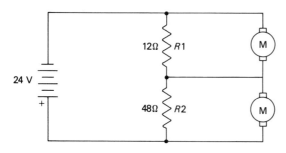

Figure 1-10. Voltage drops for series-connected motors.

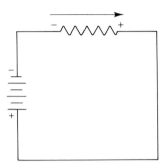

Figure 1-11. Polarity of voltage drop.

arrow points in the direction of the current. The head of the arrow is plus; its opposite end is minus.

Summary of Parallel and Series Voltage Drops

Parallel Voltage Drops

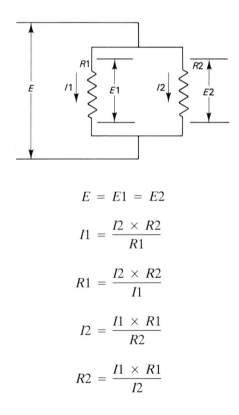

$$E = E1 = E2$$

$$I1 = \frac{I2 \times R2}{R1}$$

$$R1 = \frac{I2 \times R2}{I1}$$

$$I2 = \frac{I1 \times R1}{R2}$$

$$R2 = \frac{I1 \times R1}{I2}$$

Series Voltage Drops

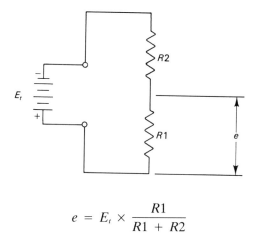

$$e = E_t \times \frac{R1}{R1 + R2}$$

where e is the voltage drop across $R1$.

Voltage Divider

$$E_t = I(R1 + R2)$$

LOADING

A motor connected to a voltage source is regarded as a load on that source. A motor that requires a large current is considered a heavy load; one that demands a small current is a light load. The terms are relative. A motor using a current of 5 A is heavier than one that operates with 1 A. But that 1-A motor is a heavy load compared to one that uses only 0.5 A.

The opposite of a loaded condition is zero load, a condition in which the motor is disconnected from its voltage source by a switch.

POWER IN DC CIRCUITS

All motors require electrical power input and supply mechanical power output. The basic unit of electrical power is the watt (W), and it can be calculated by using any one of a number of rules known as the power laws. The simplest of these laws is

$$P = E \times I$$

where

$$P = \text{the power in watts,}$$
$$E = \text{the voltage in volts, and}$$
$$I = \text{the current in amperes.}$$

If a problem involves multiples or submultiples of these basic units they must be converted before using the formula. The input power to a DC motor can be measured by a DC ammeter in the power line leading to the motor terminals and by a DC voltmeter connected across the motor's input terminals. To get a true evaluation of the input power, the motor must be running and operating its normal load. If different loads are used, that requiring the greatest input current will have the highest power input.

Power-Line Loss

The current flowing through the power line to the motor is the same anywhere along that line; the voltage at the motor input terminals may be less than that at the source. The difference between the source voltage and the motor input voltage is due to the power loss in the line. For DC motors, a DC voltmeter connected across the source voltage and a DC ammeter anywhere along the power line can supply data for calculating the power supplied by the source. The difference in power supplied at the source and that at the motor terminals is due to the loss in the line. That loss will depend on the lengths of the power lines, the DC resistance of those lines, and the amount of current flow.

DC POWER LAWS

Just as there are three different forms of Ohm's Law, so too is it possible to derive other power law arrangements. The power law as stated earlier is

$$P = E \times I$$

However, according to Ohm's Law, $E = I \times R$, and thus, since E and IR are equivalent, we can substitute IR for E in the power law.

$$P = IR \times I \quad \text{or} \quad P = I \times I \times R \quad \text{or} \quad P = I^2 R$$

It is possible to develop still another power law from Ohm's Law. Starting with the basic power law,

$$P = E \times I$$

According to Ohm's Law, however,

$$I = E/R$$

DC Power Laws

Since I and E/R are identical, one can be substituted for the other:

$$P = E \times E/R \quad \text{or} \quad P = E^2/R$$

The triangle in Figure 1-12 is a summation of the power laws for DC.

Multiples of the Watt

The watt, as in Table 1-14, has both multiples and submultiples.

Example:

An ammeter in the power line connected to a motor reads 950 mA. The voltage across the motor terminals indicates 50 V. What is the input power to the motor?

Solution:

$$\text{Power} = E \times I = 50 \times 0.950 = 47.5 \text{ W}$$

Note that the current data supplied in milliamperes was converted to amperes by moving the decimal point three places to the left. While the decimal point for the current in milliamperes is not shown, all whole numbers are assumed to end in this point.

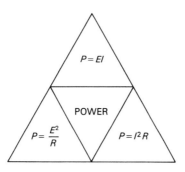

Figure 1-12. Basic power laws.

TABLE 1-14. POWER CONVERSIONS

1 watt	$= 1{,}000$ mW $= 10^3$ mW $= 10^6$ µW $= 1{,}000{,}000$ µW
1 watt	$= 0.001$ kW $= 10^{-3}$ kW
1 watt	$= 0.000001$ MW $= 10^{-6}$ MW
1 kilowatt	$= 1{,}000$ W $= 10^3$ W
1 kilowatt	$= 0.001$ MW $= 10^{-3}$ MW
1 megawatt	$= 1{,}000{,}000$ W $= 10^6$ W
1 megawatt	$= 1{,}000$ kW $= 10^3$ kW
1 milliwatt	$= 1{,}000$ µW $= 10^3$ µW
1 milliwatt	$= 0.001$ W $= 10^{-3}$ W
1 microwatt	$= 0.000001$ W $= 10^{-6}$ W
1 microwatt	$= 0.001$ mW $= 10^{-3}$ mW

Example:

A DC motor is connected to its power source by means of a line having a resistance rating of 0.04 Ω. The current flowing through the line is 840 μA or 0.000840 A. What is the voltage drop across the line and what is its power loss?

Solution: Since two wires are used to connect the motor to its source voltage, the total resistance is twice that indicated. This resistance is $0.04 \times 2 = 0.08$ Ω. The voltage drop is $E = I \times R = 840 \times 10^{-6}$ A $\times 0.08 = 0.0000672$ V. The power loss is $P = I^2R = 0.00084 \times 0.00084 \times 0.08 = 0.000000056$ W.

Note that the current data, supplied in microamperes, was converted to amperes prior to use in the formula.

The power laws and Ohm's Law formulas are related and can be used to find unknown quantities. To determine the current in amperes when the power in watts and the resistance in ohms are known, use

$$I^2 = \frac{P}{R} \quad \text{or} \quad I = \sqrt{\frac{P}{R}}$$

Example:

A 6-Ω resistor is used in series with the power line to a series-wound DC motor. The resistor dissipates 75 W in the form of heat. What is the amount of current flow through this resistor?

Solution:

$$I = \sqrt{\frac{P}{R}} = \sqrt{\frac{75}{6}}$$

$$= \sqrt{12.5} = 3.535 \text{ A}$$

To determine the resistance in ohms when the power in watts and the current in amperes are known, use

$$R = \frac{P}{I^2}$$

Example:

The working current of an armature is 8 A. It is connected to an external resistive control circuit having an operating power of 180 W. What is the value of resistance?

Solution:

$$R = \frac{P}{I^2} = \frac{180}{8 \times 8}$$

$$= 2.81 \text{ Ω}$$

DC Power Laws

To determine the power in watts when the voltage and resistance are known, use

$$P = \frac{E^2}{R}$$

Example:

A small DC motor has an armature winding whose cold resistance is 1.06 Ω. The motor is connected to a 24-V source. What is its power dissipation in watts?

Solution:

$$P = \frac{E \times E}{R} = \frac{24 \times 24}{1.06} = 543.4 \text{ W}$$

To determine the voltage when the power in watts and the resistance in ohms are known, use

$$E^2 = PR$$

or

$$E = \sqrt{PR}$$

Example:

What is the amount of DC source voltage required to drive a 500-W motor if the resistance measured across the input terminals of the motor is 15 Ω?

Solution:

$$E = \sqrt{500 \times 15}$$
$$E = \sqrt{7{,}500} = 86.6 \text{ V}$$

To determine the resistance in ohms when the power in watts and the electrical pressure in volts are known, use

$$R = \frac{E^2}{P}$$

Example:

What is the resistance of a motor's field coil that dissipates 14 W when connected to a 24-V DC source?

Solution:

$$R = \frac{E \times E}{P} = \frac{24 \times 24}{14} = \frac{576}{14}$$
$$= 41.14 \text{ Ω}$$

To determine the power in kilowatts when the voltage (in volts) and the current in amperes are known, use

$$\text{kilowatts} = \frac{\text{watts}}{1{,}000}$$

or

$$= \frac{E \times I}{1{,}000}$$

Example:

The DC voltage input to a motor is 32 V and the line current input to the motor is 15 A. What is the power input in kilowatts?

Solution:

$$\text{kW} = \frac{E \times I}{1{,}000}$$

$$= \frac{32 \times 15}{1{,}000} = \frac{480}{1{,}000} = 0.48 \text{ kW}$$

Total Power Dissipation

The total power dissipated by resistive loads, whether in series, parallel, or series-parallel circuit, is the sum of the individual powers. Consider three resistors in any circuit arrangement. Resistor R1 utilizes power P1, R2 power P2, and R3 power P3. Then

$$P_t = P1 + P2 + P3 \ldots$$

where P_t is the total power in watts. P1, P2, and P3 are also in watts.

The relationships between the Ohm's Law formulas and the DC power laws are listed in Table 1-15. When any two known values are available, unknown values for I, R, E, and P can be calculated. In this table the known values are indicated in the first column, and the unknown values can be calculated by the correct selection of a formula from one of the adjacent columns.

Summary of Power Relationships

Power

$$P = I^2 \times R$$

$$P = E^2/R$$

$$P = E \times I$$

$$\text{kW} = \frac{E \times I}{1{,}000} \quad \text{or} \quad \frac{P}{1{,}000}$$

Work

TABLE 1-15. OHM'S LAW AND POWER FORMULAS FOR DC

Known values	Formulas for determining unknown values of...			
	I	R	E	P
I & R			IR	I^2R
I & E		$\dfrac{E}{I}$		EI
I & P		$\dfrac{P}{I^2}$	$\dfrac{P}{I}$	
R & E	$\dfrac{E}{R}$			$\dfrac{E^2}{R}$
R & P	$\sqrt{\dfrac{P}{R}}$		$\sqrt{PR}$	
E & P	$\dfrac{P}{E}$	$\dfrac{E^2}{P}$		

Current

$$I = \sqrt{P/R}$$
$$I = P/E$$
$$I^2 = P/R$$

Voltage

$$E = \sqrt{P \times R}$$
$$E^2 = P \times R$$
$$E = P/I$$

Resistance

$$R = P/I \times I$$
$$R = P/I^2$$
$$R = E^2/P$$

WORK

Power used per unit of time is called *work*. This can be expressed as a formula:

$$\text{work} = P \times t$$

The basic unit of work is the watt-hour. When large amounts of power are involved, power used per unit of time is expressed in kilowatt-hours, abbreviated as kWh.

$$\text{kWh} = \frac{P \times t}{1{,}000}$$

where P is the power in watts and t is the time in hours.

Units of Work

There are three systems used in measuring the amount of work done.

    English: The units are the foot, the pound, and the second.
    CGS: The units are the centimeter, gram, and the second.
    MKS: The units are the meter, kilogram, and the second.

Work can be expressed in mechanical or electrical terms. Mechanically, work = force × distance. As an example, consider the work done by the rotating shaft of a motor. It is expressed in foot pounds. Electrically, work = power × time. Power is expressed in terms of watts multiplied by the hours during which the power is used.

The foot-pound is the work done by a force of one pound working through a distance of one foot. The kilogram-meter is the work done by a force of one kilogram working through a distance of one meter.

The erg is the work done by a force of one dyne working through a distance of one centimeter.

Gravity gives to a gram a velocity of approximately 980 centimeters per second. Therefore, it is equal to 980 dynes. If a gram weight is lifted vertically one centimeter the work done against gravity is one gram-centimeter or 980 ergs. The problem with the erg is that it is an extremely small unit and it is more convenient to use a multiple for practical measurements. The multiple employed is the joule, which is equal to 10^7 or 10,000,000 ergs.

The Time Element

It takes time to do work but the amount of work done has nothing to do with the time required. If a man weighing 150 pounds walks up all of the steps of the Washington monument, 500 feet high, he does work against gravity equal to $500 \times 150 = 75{,}000$ foot pounds, regardless of the amount of time required.

The example supplied above describes mechanical work. In electrical terms, power is the time-rate of doing work. In the English gravitational system the unit of power is the horsepower. It is the rate of doing work equal to 33,000 foot-pounds per minute or 550 foot-pounds per second.

In the CGS system the unit of power is the watt. It equals work done at the rate of one joule (10^7 ergs) per second. One horsepower is equivalent to 746 watts.

Nomograms

POWER VERSUS ENERGY

Energy is the ability to do work. The rate at which work is done is called *power*. The unit of energy or work is the *joule* (J). Power is expressed in *watts*, that is, in joules per second (J/s). In electronics, electrical power is usually measured in watts. Electrical energy can be stored, but not in resistors. Electrical energy put into a resistor is given off by that component as heat. The field and armature coils of motors that carry current produce heat from the current flow through the resistive portion of those coils. The non-resistive part of the coil, the inductive part, stores electrical energy in its magnetic field surrounding the coil. The larger the resistance of field and armature coils, and the greater the current flowing through them, the greater the amount of heat produced.

HEATING EFFECT OF A CURRENT

A current produces heat when it passes through a resistor or through any component that contains resistance as an integral part of its structure. The heat generated is proportional to the square of the current, in amperes, multiplied by the value of the resistance in ohms.

$$H = I^2 R$$

where H is in joules per second. Since joules per second = watts, the formula becomes

$$P = I^2 R$$

where

P is the power in watts,

I the current in amperes, and

R the resistance in ohms.

NOMOGRAMS

Values for resistance, current, voltage, and power can be calculated through the use of one or more selected formulas. It is an accurate method, but it involves some arithmetic. An alternative method is through the use of nomograms such as those in Figures 1-13, 1-14 and 1-15. These can be used not only to find quick answers, but also to make quick checks on formula results or to find alternative solutions. Another advantage is that no calculations are involved.

The nomogram in Figure 1-13 has three pairs of columns; in each pair, the columns are headed A and B. The use of one column A necessitates the selection

36   Basic Data   Chap. 1

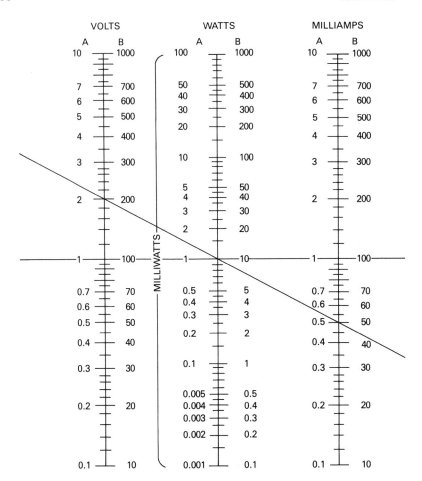

Figure 1-13. Nomogram for problems involving volts, watts, and milliamperes.

of the other A columns. Similarly, if a B column is chosen, only the other two B columns may be involved.

As in the case of formulas, two values must be known so as to find the third, unknown value. In this nomogram, the three value scales are volts, watts, and milliamperes. For power, the A column is in milliwatts, the B column in watts.

As a start, select digit 1 in all the A columns and put a ruler across each, as shown by the horizontal dashed line. This shows that with a voltage of 1 V and a current of 1 mA, the power (assuming it to be the unknown value) is 1 mW. Alternatively, either volts or watts could have been the unknowns. Unlike formulas, no conversion to basic units is required.

The ruler need not be kept horizontal as in this first example. For any problem the ruler can be placed so that the two unknown values are connected. Assume

Nomograms

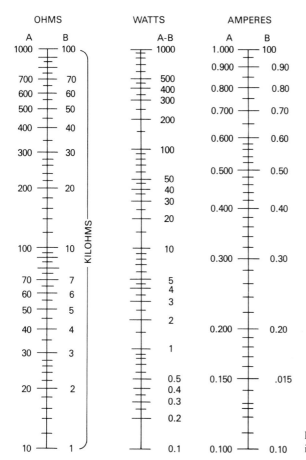

Figure 1-14. Nomogram for problems involving ohms, watts and amperes.

the voltage is 200 V and the power is 10 W. Both can be found in the B columns. The dashed line connecting these two points crosses column B at 50 mA. This is the amount of current under the conditions established in the problem.

The nomogram can also used to probe a large number of possible conditions, something that would be difficult to do with formulas. For the point that is selected, the position of the ruler will supply the other two values.

The nomogram will not supply the accuracy of formulas, and in some instances the answer will simply be a good approximation. But while formulas may be more accurate, a precision to one or two decimal places is rarely needed.

The other two nomograms, in Figures 1-14 and 1-15, are worked in a similar manner. However, Figure 1-14 has only 5 columns, due to the fact that the center column is a combined A-B type. This nomogram deals with ohms, watts, and amperes, while Figure 1-15 is used for working with ohms, volts, and watts.

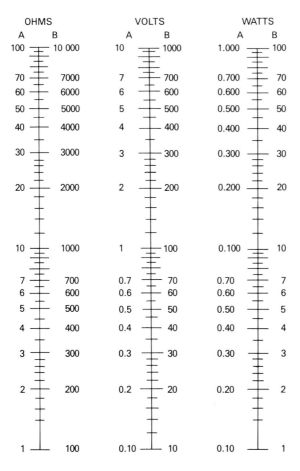

Figure 1-15. Nomogram for problems involving ohms, volts, and watts.

TEMPERATURE

In common motor practice, temperature is supplied in degrees Fahrenheit or Celsius (formerly known as centigrade). Other temperature scales, found more often in the laboratory than in industrial motor use, include Kelvin and Rankine. All of these express temperature in degrees, with the exception of the Kelvin scale. In this scale the units are in kelvins rather than in degrees. Table 1-16 supplies the formulas for making conversions between any of these temperature scales.

Example:

The ambient temperature in a room housing several motors is 40°C. What is the corresponding temperature in Fahrenheit?

Temperature

Solution:

$$°F = (°C \times 9/5) + 32$$
$$= (40 \times 9/5) + 32$$
$$= (72) + 32 = 104°F$$

TABLE 1-16. TEMPERATURE CONVERSIONS[a]

$°F = (°C \times \frac{9}{5}) + 32$

$°F = °R - 459.67$

$°F = \frac{9}{5}(K - 273.16) + 32$

$°C = \frac{5}{9}(°F - 32)$

$°C = K - 273.16$

$°C = \frac{5}{9}(°R - 491.67)$

$°R = °F + 459.67$

$°R = (°C \times \frac{9}{5}) + 491.67$

$°R = \frac{9}{5}(K - 273.16) + 491.67$

$K = °C + 273.16$

$K = \frac{5}{9}(°F - 32) + 273.16$

$K = \frac{5}{9}(°R - 491.67) + 273.16$

[a] F, Fahrenheit; C, Celsius; R, Rankine; K, Kelvins.

Temperature Conversions

There are three ways of making temperature conversions: The first and most accurate is through the use of formulas. A quicker method is by nomograms, while the third is by temperature conversion tables. The advantage of the one in Table 1-17 is that the work does not require any arithmetic. The center column, in boldface, lists numbers from -100 to $+1100$. These numbers indicate the temperature in either degrees Celsius or Fahrenheit. Locate the temperature in the center column, and the corresponding temperature, either in Celsius or Fahrenheit, is supplied in the adjoining left or right column.

TABLE 1-17. CELSIUS-FAHRENHEIT TEMPERATURE CONVERSION TABLE

−100 to 95							96 to 1100				
C		F	C		F	C		F	C		F
−73.3	**−100**	−148	6.11	**43**	109.4	35.6	**96**	204.8	304	**580**	1076
−67.8	**−90**	−130	6.67	**44**	111.2	36.1	**97**	206.6	310	**590**	1094
−62.2	**−80**	−112	7.22	**45**	113.0	36.7	**98**	208.4	316	**600**	1112
−56.7	**−70**	−94	7.78	**46**	114.8	37.2	**99**	210.2	321	**610**	1130
−51.1	**−60**	−76	8.33	**47**	116.6	37.8	**100**	212.0	327	**620**	1148
−45.6	**−50**	−58	8.89	**48**	118.4	38	**100**	212	332	**630**	1166
−40.0	**−40**	−40	9.44	**49**	120.2	43	**110**	230	338	**640**	1184
−34.4	**−30**	−22	10.0	**50**	122.0	49	**120**	248	343	**650**	1202
−28.9	**−20**	−4	10.6	**51**	123.8	54	**130**	266	349	**660**	1220
−23.3	**−10**	14	11.1	**52**	125.6	60	**140**	284	354	**670**	1238
−17.8	**0**	32	11.7	**53**	127.4	66	**150**	302	360	**680**	1256
−17.2	**1**	33.8	12.2	**54**	129.2	71	**160**	320	366	**690**	1274
−16.7	**2**	35.6	12.8	**55**	131.0	77	**170**	338	371	**700**	1292
−16.1	**3**	37.4	13.3	**56**	132.8	82	**180**	356	377	**710**	1310
−15.6	**4**	39.2	13.9	**57**	134.6	88	**190**	374	382	**720**	1328
−15.0	**5**	41.0	14.4	**58**	136.4	93	**200**	392	388	**730**	1346
−14.4	**6**	42.8	15.0	**59**	138.2	99	**210**	410	393	**740**	1364
−13.9	**7**	44.6	15.6	**60**	140.0	100	**212**	413	399	**750**	1382
−13.3	**8**	46.4	16.1	**61**	141.8	104	**220**	428	404	**760**	1400
−12.8	**9**	48.2	16.7	**62**	143.6	110	**230**	446	410	**770**	1418
−12.2	**10**	50.0	17.2	**63**	145.4	116	**240**	464	416	**780**	1436
−11.7	**11**	51.8	17.8	**64**	147.2	121	**250**	482	421	**790**	1454
−11.1	**12**	53.6	18.3	**65**	149.0	127	**260**	500	427	**800**	1472
−10.6	**13**	55.4	18.9	**66**	150.8	132	**270**	518	432	**810**	1490
−10.0	**14**	57.2	19.4	**67**	152.6	138	**280**	536	438	**820**	1508
−9.44	**15**	59.0	20.0	**68**	154.4	143	**290**	554	443	**830**	1526
−8.89	**16**	60.8	20.6	**69**	156.2	149	**300**	572	449	**840**	1544
−8.33	**17**	62.6	21.1	**70**	158.0	154	**310**	590	454	**850**	1562
−7.78	**18**	64.4	21.7	**71**	159.8	160	**320**	608	460	**860**	1580
−7.22	**19**	66.2	22.2	**72**	161.6	166	**330**	626	466	**870**	1598
−6.67	**20**	68.0	22.8	**73**	163.4	171	**340**	644	471	**880**	1616
−6.11	**21**	69.8	23.3	**74**	165.2	177	**350**	662	477	**890**	1634
−5.56	**22**	71.6	23.9	**75**	167.0	182	**360**	680	482	**900**	1652
−5.00	**23**	73.4	24.4	**76**	168.8	188	**370**	698	488	**910**	1670
−4.44	**24**	75.2	25.0	**77**	170.6	193	**380**	716	493	**920**	1688
−3.89	**25**	77.0	25.6	**78**	172.4	199	**390**	734	499	**930**	1706
−3.33	**26**	78.8	26.1	**79**	174.2	204	**400**	752	504	**940**	1724
−2.78	**27**	80.6	26.7	**80**	176.0	210	**410**	770	510	**950**	1742
−2.22	**28**	82.4	27.2	**81**	177.8	216	**420**	788	516	**960**	1760
−1.67	**29**	84.2	27.8	**82**	179.6	221	**430**	806	521	**970**	1778
−1.11	**30**	86.0	28.3	**83**	181.4	227	**440**	824	527	**980**	1796
−0.56	**31**	87.8	28.9	**84**	183.2	232	**450**	842	532	**990**	1814
−0	**32**	89.6	29.4	**85**	185.0	238	**460**	860	538	**1000**	1832
0.56	**33**	91.4	30.0	**86**	186.8	243	**470**	878	543	**1010**	1850
1.11	**34**	93.2	30.6	**87**	188.6	249	**480**	896	549	**1020**	1868
1.67	**35**	95.0	31.1	**88**	190.4	254	**490**	914	554	**1030**	1886
2.22	**36**	96.8	31.7	**89**	192.2	260	**500**	932	560	**1040**	1904
2.78	**37**	98.6	32.2	**90**	194.0	266	**510**	950	566	**1050**	1922
3.33	**38**	100.4	32.8	**91**	195.8	271	**520**	968	571	**1060**	1940
3.89	**39**	102.2	33.3	**92**	197.6	277	**530**	986	577	**1070**	1958
4.44	**40**	104.0	33.9	**93**	199.4	282	**540**	1004	582	**1080**	1976
5.00	**41**	105.8	34.4	**94**	201.2	288	**550**	1022	588	**1090**	1994
5.56	**42**	107.6	35.0	**95**	203.0	293	**560**	1040	593	**1100**	2012
						299	**570**	1058			

Temperature

Example:

What is the temperature in Fahrenheit corresponding to 42°C?

Solution: Find 42 in the center column. Move directly across to the Fahrenheit column and the answer supplied is 107.6°F.

Example:

What is the temperature in Celsius corresponding to 99°F?

Solution: Locate 99 in the center column and then move directly left to the Celsius column. The number shown is 37.2°C.

Temperature Coefficient of Resistance

The resistance of a conductor, such as the wire used in making motor armature and field coils or the wire for connecting a motor to a power source, isn't a fixed amount but depends on temperature changes. Copper is said to have a positive temperature coefficient, since its resistance increases with a rise in temperature. A positive temperature coefficient is defined as the amount of increase in resistance per ohm per degree rise in temperature above 0°C.

Some substances have a negative temperature coefficient, that is, their resistance decreases with an increase in temperature. Still others have a zero temperature coefficient, with their resistance not affected by temperature.

Copper, aluminum, and silver exhibit increases in resistance with increases in temperature, but not to the same extent. The temperature coefficient of resistance of standard annealed copper is its increase in resistance per ohm per degree rise in temperature above 0°C. The temperature coefficient of resistance of copper is 0.00427 Ω. A wire that has a resistance of 1 Ω when measured at 0°C will have a resistance of $1 + 0.00427 = 1.00427$ Ω when the temperature rises from 0°C to 1°C.

This resistance increase applies to each ohm of resistance in the copper. If the wire has an initial resistance of 5 Ω at 0°C, and the temperature rises by 1°C, then the total resistance increase will be $5 \times 0.00427 = 0.021$ Ω. The resistance of the wire at the higher temperature will be $5 + 0.021 = 5.021$ Ω.

Resistance values of copper wire are often supplied at specific temperatures, such as 0, 20, or 25°C. The temperature coefficient of resistance can be corrected to any other temperature by this equation:

$$R_{t_2} = R_{25}[1 + 0.00385(t_2 - 25)] = R_{t_1} \frac{234.5 + t_2}{234.5 + t_1}$$

where

R_{t_1} = the resistance in ohms at t_1°C,

R_{t_2} = the resistance in ohms at t_2°C, and

R_{25} = the resistance in ohms at 25°C.

Example:

A copper wire has a resistance of 45 Ω measured at 0°C. The temperature coefficient is 0.00427 Ω per degree Celsius. What is the resistance of the wire when the temperature rises to 70°F?

Solution: As a first step it will be necessary to convert degrees Fahrenheit to degrees Celsius.

$$°C = (°F - 32) \times 5/9$$
$$= (70 - 32) \times 5/9 = 38 \times 5/9 = 21.11°C.$$

Table 1-17 shows a value of 21.1°C.

A useful approach for positive temperature coefficients is to use this formula:

$$R_t = R_o(1 + at)$$

where

R_t = the resistance of the wire after the temperature increase,

R_o = the resistance of the wire at 0°C,

a = the temperature coefficient of copper (0.00427), and

t = the final temperature in degrees Celsius.

Thus

$$R_t = 45 (1 + 0.00427 \times 21.11)$$
$$= 45 (1 + 0.090139) = 45 (1.090139)$$
$$= 49.056 \, \Omega$$

WIRE

Wire is associated with motors in two ways: in the construction of the armature and field coils of motors, and in the connecting leads between the source voltage and the motors' input terminals. There is a power loss (an I^2R loss in the form of heat) in both instances.

A wire can be forced to carry more current than the amount for which it has been designed, but in the process electrical energy is converted to heat energy. Since wire has a positive temperature coefficient of resistance, this increases the resistance of the wire. The wire may also become hot enough to char or burn its insulation covering.

The current-carrying capacity of a wire is determined by its cross-sectional area. The distance between the source voltage and the motor can also be significant, since the longer the wire the greater its resistance.

The cross-sectional area of copper wire is specified by its gauge number, ranging from 0000, the thickest wire, to 46, the thinnest. Table 1-18 shows the

Wire

TABLE 1-18. WIRE TABLE[a]

Gauge number	Diameter (mils[b])	Cross section		Ohms per 1,000 ft		Ohms per mile	Pounds per 1,000 ft
		Circular mils	Square inches	25°C (= 77°F)	65°C (= 149°F)	25°C (= 77°F)	
0000	460.0	212,000.0	0.166	0.0500	0.0577	0.264	641.0
000	410.0	168,000.0	.132	.0630	.0727	.333	508.0
00	365.0	133,000.0	.105	.0795	.0917	.420	403.0
0	325.0	106,000.0	.0829	.100	.116	.528	319.0
1	289.0	83,700.0	.0657	.126	.146	.665	253.0
2	258.0	66,400.0	.0521	.159	.184	.839	201.0
3	229.0	52,600.0	.0413	.201	.232	1.061	159.0
4	204.0	41,700.0	.0328	.253	.292	1.335	126.0
5	182.0	33,100.0	.0260	.319	.369	1.685	100.0
6	162.0	26,300.0	.0206	.403	.465	2.13	79.5
7	144.0	20,800.0	.0164	.508	.586	2.68	63.0
8	128.0	16,500.0	.0130	.641	.739	3.38	50.0
9	114.0	13,100.0	.0103	.808	.932	4.27	39.6
10	102.0	10,400.0	.00815	1.02	1.18	5.38	31.4
11	91.0	8,230.0	.00647	1.28	1.48	6.75	24.9
12	81.0	6,530.0	.00513	1.62	1.87	8.55	19.8
13	72.0	5,180.0	.00407	2.04	2.36	10.77	15.7
14	64.0	4,110.0	.00323	2.58	2.97	13.62	12.4
15	57.0	3,260.0	.00256	3.25	3.75	17.16	9.86
16	51.0	2,580.0	.00203	4.09	4.73	21.6	7.82
17	45.0	2,050.0	.00161	5.16	5.96	27.2	6.20
18	40.0	1,620.0	.00128	6.51	7.51	34.4	4.92
19	36.0	1,290.0	.00101	8.21	9.48	43.3	3.90
20	32.0	1,020.0	.000802	10.4	11.9	54.9	3.09
21	28.5	810.0	.000636	13.1	15.1	69.1	2.45
22	25.3	642.0	.000505	16.5	19.0	87.1	1.94
23	22.6	509.0	.000400	20.8	24.0	109.8	1.54
24	20.1	404.0	.000317	26.2	30.2	138.3	1.22
25	17.9	320.0	.000252	33.0	38.1	174.1	0.970
26	15.9	254.0	.000200	41.6	48.0	220.0	0.769
27	14.2	202.0	.000158	52.5	60.6	277.0	0.610
28	12.6	160.0	.000126	66.2	76.4	350.0	0.484
29	11.3	127.0	.0000995	83.4	96.3	440.0	0.384
30	10.0	101.0	.0000789	105.0	121.0	554.0	0.304
31	8.9	79.7	.0000626	133.0	153.0	702.0	0.241
32	8.0	63.2	.0000496	167.0	193.0	882.0	0.191
33	7.1	50.1	.0000394	211.0	243.0	1,114.0	0.152
34	6.3	39.8	.0000312	266.0	307.0	1,404.0	0.120
35	5.6	31.5	.0000248	335.0	387.0	1,769.0	0.0954
36	5.0	25.0	.0000196	423.0	488.0	2,230.0	0.0757
37	4.5	19.8	.0000156	533.0	616.0	2,810.0	0.0600
38	4.0	15.7	.0000123	673.0	776.0	3,550.0	0.0476
39	3.5	12.5	.0000098	848.0	979.0	4,480.0	0.0377
40	3.1	9.9	.0000078	1,070.0	1,230.0	5,650.0	0.0299

[a] Standard annealed solid copper-wire table, using American Wire Gauge (AWG). Another gauge, Browne & Sharp (B & S), is identical to AWG. Diameters and cross-sectional areas are approximations.

[b] 1 mil = 0.001 inch.

relationships of bare copper wire. Known as the American Wire Gauge (AWG) the table supplies information about the resistance of wire per unit length and its resistance at a specified temperature. Figure 1-16 supplies a comparison of the cross-sectional area of some selected wires.

It is convenient to remember that the cross-sectional area of wire approximately doubles for every three gauge numbers. Thus, No. 20 wire has double the area of No. 23.

The significance here is that the current-carrying capacity is also doubled. Wires having a diameter greater than 289.0 mils (thousandths of an inch) have gauge numbers 0, 00, 000, and 0000.

Cross-Sectional Area of Wire

A wire can be round or square. In the case of square wire the cross-sectional area can be found by multiplying any two sides.

Area dimensions of wires are specified in thousandths of an inch, or mils (i.e., 1 mil is equal to 0.001 in. or 10^{-3} in.). While inches could be used, mils are more convenient, since it means working with whole numbers. A square wire that is 2 mils on a side would have $2 \times 2 = 4$ square mils area. In terms of square inches, this would be $0.002 \times 0.002 = 0.000004$ sq. in.

For round wires, the cross-sectional area can be easily calculated, since it is equal to the diameter squared. Thus,

$$A = d^2$$

where A is the cross-sectional area in circular mils and d is the diameter in mils. A circular wire having a diameter of 5 mils will have a cross-sectional area of $5 \times 5 = 25$ circular mils (Figure 1-17).

Table 1-18 supplies the diameter (in mils) and the cross-sectional area of copper wire in circular mils. To find the diameter of No. 18 wire (18 AWG) locate 18 in the gauge number column at the left. In the column to the right, the corresponding diameter in mils is supplied. Columns further to the right show the area in circular mils and the resistance in ohms per thousand feet. This resistance is measured at 77°F (25°C) and at 149°F (65°C).

Resistance of Wire

The resistance of copper wire, whether used as a field coil or an armature for a motor or as a connecting power line, is a significant physical characteristic, since it determines current-carrying capabilities. The greater the cross-sectional area of

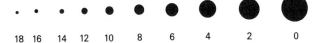

Figure 1-16. Cross-sectional areas of some copper wires from No. 18 to No. 0

Wire

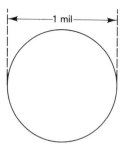

Figure 1-17. Diameter of wire is measured in mils.

a wire the more current it can handle—but at the same time the more expensive it is and the more difficult it is to work with. However, cross-sectional area isn't the only factor involved in current-carrying capability. It is also affected by the length of the wire, the ambient temperature, the temperature of the wire, and the type of insulation used.

There are two possible consequences of using the wrong gauge of wire when connecting a motor to its voltage source (assuming the wire has too high a gauge number). For one thing, because of the increased resistance there will be an increased IR (voltage) drop along the wire, thus a smaller amount of voltage available for the motor. The other is that the wire may become very hot.

The resistance of a wire is directly proportional to its length and inversely proportional to its cross-sectional area (Figure 1-18). The relationships can be conveniently expressed as

$$R = \rho \frac{L}{A}$$

where

R = the resistance in ohms,

L = the length of the wire in feet,

A = the cross-sectional area in circular mils, and

ρ = the specific resistance or the resistivity of a metal.

The metal in this case is copper. The specific resistance of copper is 10.4. Silver is 9.8, making it a better conductor than copper. Since the cross-sectional area is

Volume = length × cross-sectional area

Figure 1-18. Resistance of a conductor depends on its volume.

equal to the diameter squared, the formula for the resistance of wire can also be written as

$$R = \rho \frac{L}{d^2}$$

This formula is for wire at 20°C, in free space, and not carrying current.

Fusing Currents of Wires

Table 1-19 gives the fusing currents in amperes for five types of wires. Copper is the most common type of wire; the others are supplied for comparison. The current in amperes at which a wire will melt can be calculated from

$$I = Kd^{3/2}$$

TABLE 1-19. FUSING CURRENTS OF WIRES IN AMPERES

AWG-gauge	d (inches)	Copper $K = 10{,}244$	Aluminum $K = 7585$	German silver $K = 5230$	Iron $K = 3148$	Tin $K = 1642$
40	0.0031	1.77	1.31	0.90	0.54	0.28
38	0.0039	2.50	1.85	1.27	0.77	0.40
36	0.0050	3.62	2.68	1.85	1.11	0.58
34	0.0063	5.12	3.79	2.61	1.57	0.82
32	0.0079	7.19	5.32	3.67	2.21	1.15
30	0.0100	10.2	7.58	5.23	3.15	1.64
28	0.0126	14.4	10.7	7.39	4.45	2.32
26	0.0159	20.5	15.2	10.5	6.31	3.29
24	0.0201	29.2	21.6	14.9	8.97	4.68
22	0.0253	41.2	30.5	21.0	12.7	6.61
20	0.0319	58.4	43.2	29.8	17.9	9.36
19	0.0359	69.7	51.6	35.5	21.4	11.2
18	0.0403	82.9	61.4	42.3	25.5	13.3
17	0.0452	98.4	72.9	50.2	30.2	15.8
16	0.0508	117	86.8	59.9	36.0	18.8
15	0.0571	140	103	71.4	43.0	22.4
14	0.0641	166	123	84.9	51.1	26.6
13	0.0719	197	146	101	60.7	31.7
12	0.0808	235	174	120	72.3	37.7
11	0.0907	280	207	143	86.0	44.9
10	0.1019	333	247	170	102	53.4
9	0.1144	396	293	202	122	63.5
8	0.1285	472	349	241	145	75.6
7	0.1443	561	416	287	173	90.0
6	0.1620	668	495	341	205	107

where d is the wire diameter in inches and K has a value that depends on the metal concerned. A wide variety of factors influence the rate of heat loss, and the figures in Table 1-19 should be considered as approximations.

Current versus Wire Sizes

Table 1-20 shows the current demands of a motor and the minimum wire gauge to be used for connecting the motor to its voltage source. While the table indicates a minimum size, it is satisfactory to use a wire having a smaller gauge number. The larger the motor's current demand, the more advisable it is to make the connecting power cord as short as possible.

The wire sizes shown in Table 1-20 are only approximations and should therefore be used advisedly.

FUSES FOR MOTORS

Since the starting current of some motors is substantially higher than the running current, using an ordinary fuse will produce problems. If the starting current of a motor is 20 A and the running current is 3 A, a logical choice would be to use a 5-A fuse. Yet the 20-A starting current would blow the fuse. It is therefore better to use a time-lag type of fuse, since it will permit a high starting current but will open if the motor draws excessive current while working.

Fuse Types

There are a number of different types of fuses, as indicated in Table 1-21. While it is customary to use a single fuse, in some instances two or more are required. Thus, in a motor controlled by an electronic circuit, there may be one fuse for the electronics, or some part of it, and another fuse for the motor. Another possibility

TABLE 1-20. MOTOR CURRENT VERSUS WIRE SIZE

Amperage of motor	Minimum size wire required (AWG)
1 to 12	14
13 to 16	12
17 to 24	10
26 to 32	8
34 to 44	6
46 to 56	4
58 to 64	3
66 to 76	2
78 to 88	1
90 to 100	0

TABLE 1-21. FUSE TYPES

Time delay	Knife blade
Fast acting	Current limiting
Multi-purpose	Round semiconductor
One-time	Square semiconductor
Renewable	Midget
S-type	Medium voltage
Medium voltage	General purpose
Cartridge	Plug

is that the motor is fuse-protected in a main fuse box, with a separate fuse in the line connected to the motor. This can be the situation when not only the motor but some other appliances as well use an outlet jointly. The motor may have its own fuse, mounted on the motor frame.

Table 1-22 lists the fuse size of time-delay fuses to use for motors having various current requirements. Table 1-23 shows the fuse size of time-delay fuses to use for motors of various horsepower outputs.

Fuse Ratings

Fuses for motors for domestic use range from about 5 A to about 30 A. A common type for motors used in homes is the 15-A fuse for protecting No. 14 wire. This 15-A time-delay fuse could possibly carry a current of 20 A, provided this higher current remained for only a short time. An uninterrupted flow of 20 A would heat the fuse element, and it would open in a few minutes. For much higher currents,

TABLE 1-22. FUSE SIZE FOR MOTORS IN TERMS OF MOTOR CURRENT

Motor amperes	Maximum fuse size (in amperes)
1 to 5	15
6	20
7 to 8	25
9 to 10	30
11	35
12 to 13	40
14 to 15	45
16	50
17 to 20	60
22	70
24 to 26	80
28 to 30	90
32	100
34 to 36	110

Fuses For Motors

TABLE 1-23. FUSE SIZE FOR MOTORS IN TERMS OF HORSEPOWER OUTPUT

Horse-power	120-V Single-phase AC repulsion induction or capacitor motor (in amperes)	120-V DC compound-wound motors (in amperes)
$1/8$	4	3
$1/6$	4	3
$1/4$	6	4
$1/3$	8	6
$1/2$	10	8
$3/4$	15	10

such as 50 A or more, the fuse would open almost at once. A good rule to follow is that the fuse rating should never exceed the current-carrying capacity of the wire it protects. If the wire is designed to carry 20 A, the fuse rating should be 20 A, preferably less.

The Plug Fuse

The plug fuse (Figure 1-19) is made with a screw-in type socket and has a window at the top covered with a transparent material through which the fuse element can be seen. These fuses are sometimes available in two different shapes, hexagonal and round. If the fuse is rated at 15 A or less, the window has a hexagonal shape; fuses for more than 15 A are round. However, this isn't standard.

Plug fuses are available in current ratings of 3, 6, 12, 15, 20, 25, and 30 A. When a plug fuse blows, it may form a discoloration, a sort of black smudge against the inside of the window. The fuse rating of plug fuses is often stamped on the fuse base or may be printed on a card fitted up against the inside of the fuse window.

Whether or not a plug fuse will open depends on the amount of excess current, the time the excess current lasts, and the speed with which heat can escape. For a short-circuit condition, a fuse will open practically immediately. With an overload situation the fuse may or may not open, depending on how long the overload lasts. As a rule of thumb, if the overload is 50%, the plug fuse can open in 1 to 15 minutes. For a 10-A fuse, for example, a current of 15 A will cause it to open, not at once, but within the 15-minute period. Fuses aren't precision devices, and most of them have a 10% overload tolerance. The fact that a fuse opens in a line

Figure 1-19. Plug fuse.

connected to a motor does not always indicate motor trouble. The line may consist of a number of branches. As more loads are connected to an outlet shared by a motor, the total current drain increases, possibly enough to open the fuse.

Cartridge Fuses

The cartridge fuse (Figure 1-20) comes in two forms: renewable and non-renewable. The non-renewable type consists of a cylinder made of a hard, fiberlike material containing the fuse element, a metal filament that melts at a predetermined current value. The element is either soldered to or mechanically fastened to a pair of metal ferrules, one at each end of the fuse housing. The ferrules are actually end caps and serve as connectors when the fuse is inserted into a pair of spring metal holders.

Cartridge fuses are generally designed to handle larger currents than plug fuses, and are available in current ratings of 3, 6, 10, 25, 30, 35, 40, 50, and 60 A. Up to 30 A the fuses are 2 in. long, and for currents between 30 and 60 A they measure 3 in.

In the renewable type, the fusible metal strip can be replaced. This is done by removing the ferrules, mounting the new fuse strip, and then putting the ferrules back.

The knife-blade type of cartridge fuse, shown in Figure 1-21, is intended for heavy currents, usually from 60 to 600 A. They are larger than plug-in cartridge fuses, with those from 60 to 100 A measuring $7^{1}/_{8}$ in. in length and those with higher current ratings being longer. Knife-blade fuses are one-time fuses.

Fuse Pullers

Plug fuses can be safely removed by gripping and turning counter clockwise the insulated portion surrounding the window. While the cylindrical portion of cartridge fuses is made of an insulating material, it is easy for fingers to slip and touch the end ferrules. Since these are metal, they could conduct the current to the handler and cause an electrical shock. This can be avoided in two ways. One is to turn off the power at the main switch; another is to use a fuse puller. Figure 1-22 shows two types of fuse pullers. The one at the top contains a neon glow lamp and a pair of test leads that can be inserted in the ends of the handles of the puller. Thus this device works not only as a fuse puller, but also as a test unit. The lower drawing is simply that of a cartridge fuse puller.

Figure 1-20. Snap-in cartridge fuse.

Figure 1-21. Knife-blade cartridge fuse.

Fuses For Motors

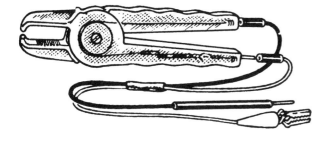

Figure 1-22. Pullers for cartridge fuses. The upper puller has a neon lamp and tests leads.

Circuit Breakers

A circuit breaker is a combined fuse and switch. Unlike plug- and cartridge-type fuses, no replacement is needed. When a circuit breaker opens, equivalent to a blown fuse, all that is needed is to reset the breaker.

Circuit breakers are available with a time-lag feature so that they can also be used with motors having a high starting current. Like fuses, circuit breakers are constructed in various ways. They may be heat-operated (thermal) or magnetically operated, or may work through a combination of these properties.

The thermal breaker has a bimetallic element. This is an element made by bonding two different metals, with each of the metals having a different temperature coefficient of expansion. The two metals are affected differently by increases in temperature, with one expanding more rapidly than the other. Since the two metals are joined, a rise in temperature will make the element bend. The bimetallic element will then act as a latch, tripping the circuit breaker and causing it to open. The bimetallic element will do this when the current exceeds a predetermined value.

Once the breaker is open, current can no longer flow. The breaker works as a switch. The switch, whose external structure is somewhat like an ordinary light switch, can then be reset to its closed position. Current will once again pass through the bimetallic element. If the condition that caused the excessive current flow is removed, the circuit breaker will remain closed.

A magnetic circuit breaker, as its name implies, works as a magnet. Whenever an electrical current flows through a coil, it becomes an electromagnet, capable of attracting and holding a bit of ferrous metal. This metal could be part of a switch.

As long as current through the magnetic circuit breaker coil remains normal, the magnet isn't strong enough to pull the metal strip away from its closed-switch position. However, in case of an overload or a short circuit, the current through the magnetic circuit-breaker coil increases substantially. The coil then becomes a much stronger magnet and attracts the metal strip to itself. This metal strip is part of a switch; in moving toward the magnet, the switch opens the power line connected to the motor.

Figure 1-23 shows a small circuit breaker box; some contain many more breakers. The breaker at the top is the main unit for controlling all currents. The four other breakers are intended for separate power lines.

Plug-to-Circuit-Breaker Conversions

Circuit breakers are much more convenient than other types of fuses. An easy method for converting to the more convenient type is to substitute mini–circuit breakers for plug fuses. These look like plug fuses (Figure 1-24) and fit directly as replacements. They have a reset feature consisting of a small pushbutton in the center of the fuse. When the fuse blows, it can be reset by pushing the button in.

Circuit breakers used in industrial applications are constructed somewhat differently than home types. Because of the higher inductance of the holding coil, the rapid opening may induce a high voltage, causing the circuit breaker to arc.

Figure 1-23. Small circuit-breaker box.

Horsepower 53

Figure 1-24. Component converts fuse box into circuit breaker type.

Air-type circuit breakers dissipate this arc through carbon electrodes. Large breakers may be immersed in oil to counteract the effects of the arc.

HORSEPOWER

The input to a motor is measured in terms of electrical power, that is, in watts, in a multiple such as kilowatts, or in some submultiple such as milliwatts or microwatts. The output of the motor is expressed in mechanical terms of horsepower (hp). This is the basic unit, and there are no specific submultiples or multiples. Values of less than 1 hp are shown in fractions, such as 1/4, 1/2, and so on, or less often by a decimal.

There is a relationship between electrical power and horsepower based on the number of input watts required to produce horsepower output: 1 hp is equivalent to 745.7 W. This number, 745.7, is generally rounded to 746.

Table 1-24 supplies power input in watts and the equivalent amount of horsepower for values ranging from 1 to 100 W. The same table can be expanded by moving the decimal point either left or right. Thus, 1 W is equivalent to 0.001341 hp. Moving the decimal point one place to the right converts 1 W to 10 W and horsepower to 0.01341. The decimal point can be moved two or more places to the right for both watts and horsepower.

For small values of electrical power move the decimal point one or more places to the left. Thus, 0.01 W then has an equivalent mechanical power of 0.00001341 hp.

Table 1-25 shows the equivalent value in watts when the horsepower is known. Thus, 0.01 hp corresponds to 7.457 W. As in the case of the preceding table, the decimal point can be moved left or right.

Example:

A small motor is rated at 0.18 hp. What is its rating in watts?

Solution: Locate 0.18 hp in Table 1-25. The wattage rating is shown at its right: 134.226 W.

Example:

A motor's rating on its data plate is 31 W. What is its corresponding horsepower equivalent?

TABLE 1-24. WATTS VERSUS HORSEPOWER

Watts	Horsepower	Watts	Horsepower	Watts	Horsepower
1	0.001341	34	0.045594	67	0.089847
2	.002682	35	.046935	68	.091188
3	.004023	36	.048278	69	.092529
4	.005364	37	.049617	70	.093870
5	.006705	38	.050958	71	.095211
6	.008046	39	.052299	72	.096552
7	.009387	40	.053640	73	.097893
8	.010728	41	.054981	74	.099234
9	.012069	42	.056322	75	.100575
10	.013410	43	.057663	76	.101916
11	.014751	44	.059004	77	.103257
12	.016092	45	.060345	78	.104598
13	.017433	46	.061686	79	.105939
14	.018774	47	.063027	80	.107280
15	.020115	48	.064368	81	.108261
16	.021456	49	.065709	82	.109962
17	.022797	50	.067050	83	.111303
18	.024138	51	.068391	84	.112644
19	.025479	52	.069732	85	.113985
20	.026820	53	.071073	86	.115326
21	.028161	54	.072414	87	.116667
22	.029502	55	.073755	88	.118008
23	.030843	56	.075096	89	.119349
24	.032184	57	.076437	90	.120690
25	.033525	58	.077778	91	.122031
26	.034866	59	.079119	92	.123372
27	.036207	60	.080460	93	.124713
28	.037548	61	.081801	94	.126054
29	.038889	62	.083142	95	.127395
30	.040230	63	.084483	96	.128736
31	.041571	64	.085824	97	.130077
32	.042912	65	.087165	98	.131418
33	.044253	66	.088506	99	.132759
				100	.134100

Solution: Find 31 in the left column in Table 1-24. This corresponds to 0.041571 hp in the column to the right.

Another advantage of both tables is that they can be used to verify the accuracy of solutions to problems involving electrical and mechanical power.

Horsepower can be expressed not only in terms of watts but in foot-pounds per second (of which more later) and kilowatts as well. Thus,

$$1 \text{ hp} = 746 \text{ W} = 550 \text{ ft-lb/s}$$

TABLE 1-25. HORSEPOWER VERSUS WATTS

Horsepower	Watts	Horsepower	Watts	Horsepower	Watts
0.01	7.457	0.34	253.538	0.67	499.619
.02	14.914	.35	260.995	.68	507.076
.03	22.371	.36	268.452	.69	514.533
.04	29.828	.37	275.909	.70	521.990
.05	37.285	.38	283.366	.71	529.447
.06	44.742	.39	290.823	.72	536.904
.07	52.199	.40	298.280	.73	544.361
.08	59.656	.41	305.737	.74	551.818
.09	67.113	.42	313.194	.75	559.275
.10	74.570	.43	320.651	.76	556.732
.11	82.027	.44	328.108	.77	574.189
.12	89.484	.45	335.565	.78	581.646
.13	96.941	.46	343.022	.79	589.103
.14	104.398	.47	350.479	.80	596.560
.15	111.855	.48	357.936	.81	604.017
.16	119.312	.49	365.393	.82	611.474
.17	126.769	.50	372.850	.83	618.931
.18	134.226	.51	380.307	.84	626.388
.19	141.683	.52	387.764	.85	633.845
.20	149.140	.53	395.221	.86	641.302
.21	156.597	.54	402.678	.87	648.759
.22	164.054	.55	410.135	.88	656.216
.23	171.511	.56	417.592	.89	663.673
.24	178.968	.57	425.049	.90	671.130
.25	186.425	.58	432.506	.91	678.587
.26	193.882	.59	439.963	.92	686.044
.27	201.339	.60	447.420	.93	693.501
.28	208.796	.61	454.877	.94	700.958
.29	216.253	.62	462.334	.95	708.415
.30	223.710	.63	469.791	.96	715.872
.31	231.167	.64	477.248	.97	723.329
.32	238.624	.65	484.705	.98	730.786
.33	246.081	.66	492.162	.99	738.243
				1.00	745.700

and in terms of kilowatts,

$$hp = \frac{kW \times 1{,}000}{746}$$

These formulas can be transposed in terms of power in watts or kilowatts:

$$1 \text{ W} = 1/746 \text{ hp} = 0.00134 \text{ hp}$$

$$1 \text{ kW} = \frac{hp \times 746}{1{,}000}$$

Summary of Formulas Involving Horsepower

$$1 \text{ hp} = 746 \text{ W} = 550 \text{ ft-lb/s}$$

$$1 \text{ W} = 1/746 \text{ hp} = 0.001341 \text{ hp}$$

$$kW = \frac{hp \times 746}{10^3}$$

$$hp = \frac{\text{ft-lb/min.}}{33{,}000}$$

$$hp = \frac{E \times I}{746}$$

where

E = voltage in volts, and

I = current in amperes.

Current (DC)

$$I = \frac{746 \times hp}{E \times \text{eff}}$$

where

eff = efficiency.

Current (AC)

$$I = \frac{746 \times hp}{E \times \text{eff} \times \text{pf} \times 1.732}$$

where pf = power factor.

THE FOOT-POUND

The ultimate characteristic of a motor is its ability to do work. A motor can operate a printing press, a lathe, a drill, a saw, and so on. The amount of work done depends on the force that is used and the distance through which this force acts. This concept is expressed in terms of a formula:

$$W = F \times S$$

where

W = work,

F = force, and

S = distance.

The Foot-Pound

In the English system of measurements, if a pound of force is applied through a distance of 1 ft, the result is 1 foot-pound (ft-lb) of work. Both the force and the distance must have numerical values greater than zero. If either F or $S = 0$, no work is done. The foot-pound is different from the unit of torque, the pound-foot. For every 33,000 foot-pounds of work per minute, 1 hp is produced. To change foot-pounds per minute into horsepower, divide by 33,000.

Because of the relationship between electrical power in watts and mechanical power in horsepower, both are involved in work. Electrical power used per unit of time can be given as

$$\text{Work} = P \times t$$

where P is the power in watts and t is the time in hours. This results in a basic unit known as the watt-hour. As mentioned earlier, when large amounts of power are involved, power used per unit of time is expressed in kilowatt-hours, abbreviated as kWh.

Power is the rate at which work is done and can be supplied in a formula derived from work $= P \times t$. Thus,

$$\text{Power} = \frac{\text{work}}{\text{time}}$$

Table 1-26 shows mechanical and electrical power equivalents.

The data shown in Table 1-27 represent the approximate full-load currents for DC motors having line inputs of 120, 240, and 550 V. These currents are with the motors operating at full load and represent average values. There can be current variations from the figures supplied here, with changes caused by differences in speed and load.

The current requirement of a 240-V DC motor can also be calculated by

$$\text{Current} = \frac{\text{hp} \times 746}{\text{Voltage} \times \text{Eff.}}$$

The approximate rule for 240-V motors is 4 A per 1 hp.

Mechanical-power-to-electrical-power conversions, and vice versa, can be performed by consulting a conversion table such as that in Table 1-28.

TABLE 1-26. MECHANICAL AND ELECTRICAL POWER EQUIVALENTS

1 horsepower	= 33,000 ft-lb/min.
1 horsepower	= 550 ft-lb/s
1 horsepower	= 0.7457 kW
1 kilowatt	= 1,000 J/s
1 kilowatt	= 1.341 hp
1 kilowatt	= 44,250 ft-lb/min.
1 kilowatt	= 737.5 ft-lb/s

TABLE 1-27. DC MOTOR FULL-LOAD CURRENTS

Horsepower of motor	DC motors		
	120 V	240 V	550 V
1/4	—	—	—
1/2	4.5	2.3	—
3/4	6.5	3.3	1.4
1	8.4	4.2	1.7
1½	12.5	6.3	2.6
2	16.1	8.3	3.4
3	23.0	12.3	5.0
5	40.0	19.8	8.2
7½	58.0	28.7	12.0
10	75.0	38.0	16.0
15	112.0	56.0	23.0
20	140.0	74.0	30.0
25	185.0	92.0	38.0
30	220.0	110.0	45.0
40	294.0	146.0	61.0
50	364.0	180.0	75.0
60	436.0	215.0	90.0
75	540.0	268.0	111.0
100	—	357.0	146.0
125	—	443.0	184.0
150	—	—	220.0
175	—	—	—
200	—	—	295.0

METRIC AND ENGLISH UNITS

Dimensions used in working with motor systems can be in the English system—inches (in.), feet (ft), yards (yd), and miles (mi)—or in the metric system—millimeters (mm), centimeters (cm), meters (m), and kilometers (km). A kilometer is a thousand meters; a millimeter is one-thousandth of a meter; and a centimeter is one-hundredth of a meter. In a few rare instances, the two systems are intermixed. Table 1-29 shows some of the basic relationships.

In one application the resistance of wire per unit length may be supplied in feet, yards, meters, and kilometers. Table 1-30 supplies conversion factors for lengths in the two systems.

In some instances the conversion is simply from meters to feet: Table 1-31 can be used for this. To convert from feet to meters, use Table 1-32.

Metric and English Units

TABLE 1-28. POWER CONVERSIONS

To convert from	To	Multiply by
kilowatt-hours	horsepower-hours	1.3410
watts	horsepower	1.3410×10^{-3}
foot-pounds	kilowatt-hours	3.766×10^{-7}
foot-pounds per minute	kilowatts	2.260×10^{-5}
foot-pounds per second	kilowatts	1.356×10^{-3}
horsepower	kilowatts	0.7457
watts	joules per second	1.0
kilowatts	horsepower	1.3410
kilowatts	joules per second	1,000
foot-pounds per second	horsepower	1.8182×10^{-3}
foot-pounds per second	joules per second	1.3558
foot-pounds per minute	horsepower	3.0303×10^{-6}
foot-pounds per minute	joules per second	0.02260
joules per second	horsepower	1.341×10^{-3}
horsepower-hours	kilowatt-hours	0.7457
horsepower-hours	watt-hours	745.7
kilowatt-hours	horsepower-hours	1.3410
kilowatt-hours	watt-hours	1,000
watt-hours	horsepower-hours	1.3410×10^{-3}

TABLE 1-29. METRIC AND ENGLISH UNITS

1 inch = 2.54 cm
1 inch = 25.40 mm
1 inch = .0254 m
1 foot = 304.80 mm
1 foot = 30.48 cm
1 centimeter = 0.3936 in.
1 centimeter = .03280 ft
1 millimeter = .0393 in.
1 millimeter = .00328 ft
1 meter = 39.3696 in.
1 meter = 3.2808 ft

TABLE 1-30. RESISTANCE CONVERSIONS IN ENGLISH AND METRIC

From	To	Multiply by	From	To	Multiply by
ohms per foot	ohms per meter	0.3048	ohms per meter	ohms per foot	3.2808
ohms per kilometer	ohms per 1000 ft	0.3048	ohms per 1000 ft	ohms per kilometer	3.2808
ohms per kilometer	ohms per 1000 yd	0.9144	ohms per 1000 yd	ohms per kilometer	1.0936

TABLE 1-31. METERS TO FEET

Meters	Feet	Meters	Feet	Meters	Feet	Meters	Feet
1	3.2808	26	85.302	51	167.32	76	249.34
2	6.5617	27	88.583	52	170.60	77	256.62
3	9.8425	28	91.863	53	173.88	78	255.90
4	13.123	29	95.144	54	177.16	79	259.19
5	16.404	30	98.425	55	180.45	80	262.47
6	19.685	31	101.71	56	183.73	81	265.75
7	22.966	32	104.99	57	187.01	82	269.03
8	26.247	33	108.27	58	190.29	83	272.31
9	29.527	34	111.55	59	193.57	84	275.59
10	32.808	35	114.83	60	196.85	85	278.87
11	36.089	36	118.11	61	200.13	86	282.15
12	39.370	37	121.39	62	203.41	87	285.43
13	42.651	38	124.67	63	206.69	88	288.71
14	45.932	39	127.95	64	209.97	89	291.99
15	49.212	40	131.23	65	213.25	90	295.27
16	52.493	41	134.51	66	216.53	91	298.56
17	55.774	42	137.80	67	219.82	92	301.84
18	59.055	43	141.08	68	223.10	93	305.12
19	62.336	44	144.36	69	226.38	94	308.40
20	65.617	45	147.64	70	229.66	95	311.68
21	68.897	46	150.92	71	232.94	96	314.96
22	72.178	47	154.20	72	236.22	97	318.23
23	75.459	48	157.48	73	239.50	98	321.52
24	78.740	49	160.76	74	242.78	99	324.80
25	82.021	50	164.04	75	246.06	100	328.08

TORQUE

Torque, expressed in pound-feet (lb-ft), is the twisting or turning power of the motor shaft. It is proportional to the current and the density of the magnetic fields of the armature and field coils. Torque is based on the product of NI, where N is the number of coil turns and I is the current in amperes.

Pound-Feet

Torque can be determined by

$$T = F \times R$$

where

T = the torque in pound-feet,

R = the radius through which the force acts, and

F = the force.

Torque

TABLE 1-32. FEET TO METERS

Feet	Meters	Feet	Meters	Feet	Meters	Feet	Meters
1	0.3048	26	7.9248	51	15.545	76	23.165
2	0.6096	27	8.2296	52	15.850	77	23.470
3	0.91440	28	8.5344	53	16.154	78	23.774
4	1.2192	29	8.8392	54	16.459	79	24.079
5	1.5240	30	9.1440	55	16.764	80	24.384
6	1.8288	31	9.4488	56	17.069	81	24.689
7	2.1336	32	9.7536	57	17.374	82	24.994
8	2.4384	33	10.058	58	17.678	83	25.298
9	2.7432	34	10.363	59	17.983	84	25.603
10	3.0480	35	10.668	60	18.288	85	25.908
11	3.3528	36	10.973	61	18.593	86	26.213
12	3.6576	37	11.278	62	18.898	87	26.518
13	3.9624	38	11.582	63	19.202	88	26.822
14	4.2672	39	11.887	64	19.507	89	27.127
15	4.5720	40	12.192	65	19.812	90	27.432
16	4.8768	41	12.497	66	20.117	91	27.737
17	5.1816	42	12.802	67	20.422	92	28.042
18	5.4864	43	13.106	68	20.726	93	28.346
19	5.7912	44	13.411	69	21.031	94	28.651
20	6.0960	45	13.716	70	21.336	95	28.956
21	6.4008	46	14.021	71	21.641	96	29.261
22	6.7056	47	14.326	72	21.946	97	29.566
23	7.0104	48	14.630	73	22.250	98	29.870
24	7.3152	49	14.935	74	22.555	99	30.175
25	7.6200	50	15.240	75	22.860	100	30.480

In terms of horsepower,

$$\text{hp} = \frac{TN}{5252}$$

and

$$T = \frac{5252 \text{ hp}}{N}$$

where

T = the torque in pound-feet,

N = the revolutions per minute (rpm), and

hp = the horsepower.

Example:

A 15-hp motor operates at a speed of 900 rpm. What is the amount of torque this motor will supply to its load?

Solution:

$$T = \frac{5252 \text{ hp}}{N} = \frac{5252 \times 15}{900} = 87.53 \text{ lb-ft}$$

The relationship between horsepower and torque is shown in these two generalized formulas:

$$(\text{hp}) = \frac{P \times 2\pi \times R \times N}{33,000 \times 12}$$

$$(\text{hp}) = \frac{P \times R \times N}{63,000} \quad \text{or} \quad \frac{T \times N}{63,000}$$

where

P = the load in pounds,

π = 3.1416,

R = the distance in inches,

N = the revolutions per minute, and

T = the torque.

The fact that a motor develops torque doesn't necessarily mean the object to which the turning force of the shaft is applied will move. The torque may be inadequate for the load. A large torque may exist and yet not a single pound-foot of work may be done.

EFFICIENCY

Assuming a perfect motor, that is, one completely without losses, an electrical input of 746 W would result in 1 hp at its output shaft. However, there are losses in the form of heat and losses resulting from eddy currents, friction, magnetization, and windage. Consequently, while the input may be 746 W, the mechanical energy output will be less than 1 hp.

The efficiency of a motor is a measure of its ability to minimize its losses. The efficiency of a motor can be written as

$$\text{Efficiency} = \frac{\text{output}}{\text{input}}$$

This statement can also be written as

$$\text{Efficiency} = \frac{\text{input} - \text{losses}}{\text{input}}$$

In the formula for efficiency, the output is always less than the input, hence the result is always smaller than 1. The result will be a decimal, such as 0.90, 0.85,

Efficiency

and so on. It can be converted to a whole number by multiplying it by 100 and in that case is referred to as the *percentage of efficiency*. If the efficiency of a motor is 0.93, it is described as being 93% efficient. A percentage cannot be used in solving motor problems in the formula unless it is first converted to a decimal by dividing it by 100.

Summary of Efficiency Formulas

$$\text{Percent efficiency} = \frac{\text{output}}{\text{input}} \times 100$$

$$\text{Input} - \text{output} = \text{losses}$$

$$\text{Input} - \text{losses} = \text{output}$$

$$\text{Output} + \text{losses} = \text{input}$$

$$\text{Efficiency} = \frac{\text{input} - \text{losses}}{\text{input}}$$

$$\text{Efficiency} = \frac{\text{output}}{\text{output} + \text{losses}}$$

$$\text{Efficiency} = \frac{\text{output power} \times \text{time}}{\text{input power} \times \text{time}}$$

Since the times of input power and output power are the same, these two terms cancel.

Overall Efficiency

The efficiency formulas supplied are for motors only. The overall efficiency is the product of the motor efficiency and the component it is driving.

$$\text{Overall efficiency} = \text{motor efficiency} \times \text{load efficiency}$$

If a motor has an efficiency of 85% and its load has an efficiency of 75%, the overall or system efficiency is $0.85 \times 0.75 = 0.6375$, or 63.75%.

For all DC motors horsepower output can be supplied as

$$\text{hp} = \frac{\eta I E}{746}$$

where

hp = the horsepower output,

η = the efficiency,

I = the line current, and

E = the line voltage.

The efficiency factor can be combined with formulas involving either power in terms of watts or horsepower. The formula for horsepower can be stated as

$$\text{hp} = \frac{E \times I}{746}$$

Transposing:

$$746 \times \text{hp} = E \times I$$

$$I = \frac{746 \times \text{hp}}{E}$$

If the efficiency is less than 100%, the formula becomes

$$I = \frac{746 \times \text{hp}}{E \times \eta}$$

The losses sustained in an electric motor can be determined by calculating the output horsepower, converting it to watts, and then comparing it to the input power, also in watts. Alternatively, the input power can be converted to horsepower. The motor loss can be determined by subtracting the horsepower output from the horsepower input. The answer will be in horsepower, but this can then be converted to electrical power in watts.

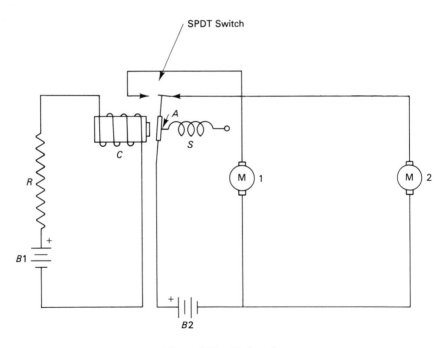

Figure 1-25. Basic relay.

Relays

Example:

A 4-hp motor requires an input of 4 kW when operating at its full load. What is the efficiency of this motor?

Solution: The output is $4 \times 746 = 2984$ W. The input is $4 \times 1{,}000 = 4{,}000$ W. The efficiency is $2{,}984/4{,}000 = 0.7460 = 74.60\%$.

RELAYS

A relay can be regarded as a limited-motion motor. Like the motor, it works on the development of a magnetic field, using that magnetic field for attracting a small length of iron or steel, pivoted at one end. Known as an armature, this length of metal works as a switch, producing results similar to mechanical switches.

TABLE 1-33. BASIC RELAY CIRCUITS

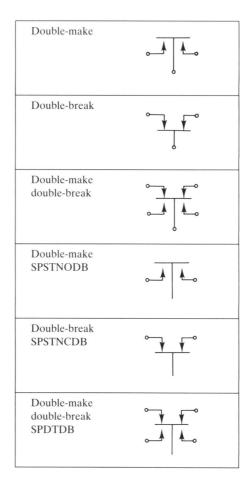

Make SPSTNO	
Break SPSTNC	
Break, make SPDT	
Make before break	
Break, make, break	
Single-pole, double-throw, SPDTNO	
Double-make	
Double-break	
Double-make double-break	
Double-make SPSTNODB	
Double-break SPSTNCDB	
Double-make double-break SPDTDB	

Figure 1-25 shows the basic structure. A current from B1 flows through an iron-core coil (C) and makes it into an electromagnet. The magnetic field around this coil attracts armature A so that it makes contact with the left terminal of the single-pole, double-throw switch. This completes the circuit to motor 1, and this motor starts working.

A mechanical switch (not shown in the drawing) can interrupt the current flow to the relay coil and demagnetize it. The spring, S, now pulls the armature to the second contact. This closes the circuit to motor 2, and its armature begins rotation. With the help of the relay, either one motor or the other can function.

The relay switch in Figure 1-25 is a single-pole double-throw type (SPDT) type, but a number of variations are possible, including single-pole single-throw (SPST) types. Several variations are shown in Table 1-33.

A relay switch can be normally open (NO) or normally closed (NC). A switch that is normally open is a "make" type, that is, the switch contacts must meet or "make." A switch that is normally closed is a "break" type, that is, the switch contacts must open or "break."

Relays can be used to control motors that are at some distance from an operating point or that must work in a hazardous environment. They can be used to turn high voltages on or off when these are the source voltages for motors. They can be used to turn motor control circuits on or off. They can be used to protect a motor against phase failure of power or against phase reversal. Phase reversal can cause a motor to turn in an opposite direction, a condition that could be hazardous in the case of elevators, hoists, and cranes.

chapter two

DC Power Sources

There are three types of power sources for DC motors: batteries, electronic power supplies, and motor generators. Batteries are commonly used in the home, electronic power supplies in the home and industry, and motor generators in industry.

BATTERIES

There are numerous types of batteries, both renewable and non-renewable. A renewable battery, also called a secondary battery, is one that can be recharged; a non-renewable, also known as a primary, is disposable. Renewables have a much longer working life, but a higher initial cost.

As a power source, batteries supply voltage and current. The fact that a meter, especially one that has a very high resistance, indicates full voltage when connected across the terminals of a battery, is no indication that the battery is charged. This terminal voltage is significant only if the measurement is made when the battery is connected and is operating its usual full load.

Physical Dimensions

Outline drawings showing dimension details are helpful when the size of the battery must be considered. The drawings in Figure 2-1 shows the measurements of AA, C, D, and 9-V batteries.

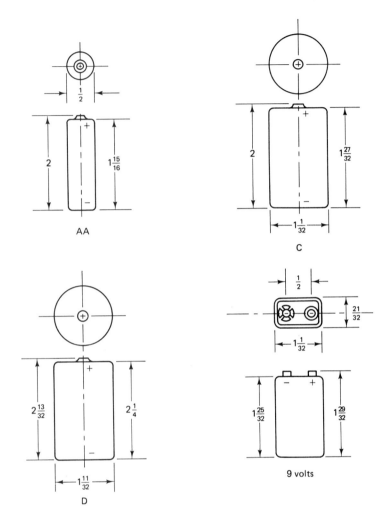

Figure 2-1. Physical data and outlines of AA, C, D, and 9-V cells (dimensions in inches).

Cells in Series

Technically, a *battery* consists of two or more cells, but in practice the term refers as often to a single cell. To obtain more voltage than that supplied by a single cell, two or more can be connected in series aiding (Figure 2-2). The total voltage supplied by using this type of connection is

$$E_t = E1 + E2$$

where E_t is the total voltage measured from the minus terminal of one battery to the plus terminal of the second. $E1$ and $E2$ represent the two cells, but any number of identical cells can be used.

Batteries

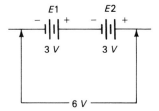

Figure 2-2. Cells in series aiding.

This formula, revised for three or more cells in series, is expressed as

$$E_t = E1 + E2 + E3 + E_n$$

In this formula E_n indicates an indeterminate number of cells. Cells connected in series should be as alike and as similarly charged as possible. Though the total voltage is the sum of the individual voltages, the current capability is that of the weakest single cell in the group. Series connections are used to meet the input voltage requirement of a motor.

Cells in Parallel

Two or more cells can be connected in parallel, as indicated by the circuit in Figure 2-3. The total available voltage is that of any one of the cells, assuming they are all identical, as indeed they should be. The total available current is the total current capacity of all of the cells

$$E_t = E1 = E2 \quad \text{or} \quad E2 = E1$$

Cells in Series-Parallel

To obtain a higher battery voltage and current, cells can be wired in series-parallel, as indicated in the circuit in Figure 2-4. The configuration is expressed as

$$E_t = E1 + E2 + (E3 \text{ or } E4)$$

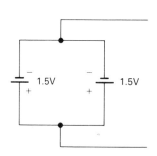
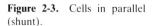
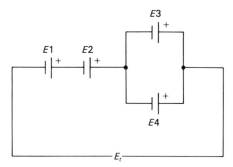

Figure 2-3. Cells in parallel (shunt).

Figure 2-4. Cells in series-parallel.

Cell Polarity

The polarity of a cell is a reference to its positive (+) and negative (−) terminals. These terminals can both be located on the top of the cell, or one on top and the other on the bottom, depending on the physical construction of the cell.

Cells in Series Opposing

Cells can be connected so as to have their voltages in opposition to each other (Figure 2-5). If both cells have the same terminal voltage, the total voltage as measured across both cells will be zero and the cells will not supply an output current. If the voltages are not the same, the smaller is subtracted from the larger. A circuit arrangement of this kind is sometimes used in electronic circuits for motor control.

The opposition voltage concept is also applicable to resistors that have a current flowing through them (Figure 2-6). This is a more practical approach than using batteries, since the current flow, hence the voltage drops, can be more precisely controlled.

AMPERE-HOUR RATINGS

The amount of current that can be delivered by a battery, assuming the battery is charged, is the product of its current capability and the length of time during which that current is being delivered to a motor. It can be stated as

$$Ah = I \times t$$

where

Ah = ampere-hours,

I = the current in amperes, and

t = the time in hours.

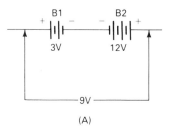

(A)

Figure 2-5. Cells in series opposing.

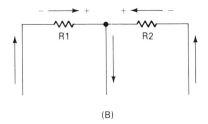

(B)

Figure 2-6. Resistor arrangements for IR drops in opposition.

Battery Types and Their Characteristics

If the current flow is in milliamperes or microamperes, these must first be converted to amperes, or else the battery must be rated in milliampere- or microampere-hours.

Example:

The field coil in a motor has a resistance of 6Ω and operates from a 12-V battery source. The motor is allowed to run for 2 min. What is the ampere-hour rating of the battery?

Solution: The current must first be calculated using Ohm's Law: $I = E/R = 12/6 = 2$ A. The time must also be converted from minutes to hours: 2 min. = 2/60 hr = 1/30 hr. The ampere-hour rating of the voltage source is $Ah = I \times t = 2 \times 1/30 = 2/30 = 1/15 = 0.0667$ ampere-hour.

Ampere-hours can also be rated in smaller units. One of these is the milliampere-hour (mAh) and is the amount of current, measured in milliamperes, that can be supplied by a battery in 1 hr. A microampere-hour (μAh) is the total current available in microamperes in 1 hr.

REFERENCE POINTS

In motor systems, ground is often used as a reference point or may be a connecting point for power leads going into a motor's power input terminals. Ground can be the earth or some metal (such as a water pipe) going into the ground. In some instances ground may consist of a large area metal plate buried in the earth. In a motor circuit ground may represent a common connecting point.

Voltage Reference Points

While ground, a common bus, or a neutral wire are commonly used as voltage reference points, any voltage point can also act as a reference. In Figure 2-7(a), point A is 12 V negative with respect to point B. Point B is 12 V positive with respect to point A. In Figure 2-7(b), point A is 6 V negative with respect to ground. Point B is 6 V positive with respect to ground. The voltage from A to B is 12 V. In Figure 2-7(c), both points A and B are 6 V positive with respect to ground. The voltage between points A and B is zero. In Figure 2-7(d), points A and B are both 6 V negative with respect to ground. The voltage between points A and B is zero.

BATTERY TYPES AND THEIR CHARACTERISTICS

There are numerous types of batteries, and the selection of a specific battery type depends on the motor's voltage and current requirements, the physical size of the battery and the space available for its use, the voltage and current capability of the battery (its rating in ampere-hours), whether it is rechargeable, and its cost.

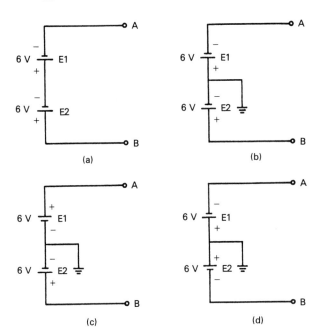

Figure 2-7. Voltage reference points.

In battery-operated motors it is seldom possible to substitute one battery type for another because of the differences in physical and electrical characteristics.

Battery types include the lead-acid, nickel-cadmium, zinc-carbon, alkaline, mercuric oxide, lithium, silver oxide, zinc chloride, and zinc-air. Every battery can be considered as a multifunction type and thus usage isn't restricted.

Battery Identification

There are various ways of identifying batteries. While all batteries have a possible use for powering a motor (depending on the motor's voltage and current requirements), many were made for some other specific purpose. Thus the 9-V battery is widely used for powering transistor radios and is often referred to as a transistor radio battery, but this does not exclude its use for motors.

Various letters are also used, such as AA, AAA, and D (Figure 2-8). Some batteries are made for applications where space limitation or weight is important, such as in hearing aids and pagers. In some instances the chemicals used in making the battery help in its identification. The lead-acid battery is sometimes called an auto battery, though it has many other uses, such as powering electric motors.

Manufacturers may assign their own part numbers to batteries, but there is no industry-wide standardization. Thus one manufacturer's part number may have no relationship to some other manufacturer's part number, even though the batteries may be interchangeable and are identical as far as voltage and current are concerned. As an aid to making a selection, batteries are sometimes cross-referenced in manufacturers' literature.

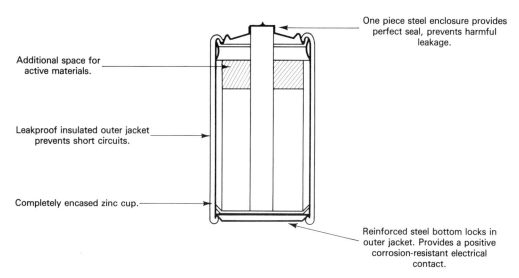

Figure 2-8. Size D cell.

Still another method of recognizing specific battery types is through alphanumeric designations supplied by various agencies. One of these is the American National Standards Institute (ANSI) in New York. Another is the International Electrotechnical Commission (IEC) in Geneva, Switzerland. IEC is represented in the U.S. by a special committee that works with ANSI for the establishment of component standards. Battery manufacturers sometimes use an ANSI designation only, and sometimes both ANSI and IEC.

There are two rechargeable battery types: the lead-acid and the nickel-cadmium, the latter popularly known as the "nicad." All the others are disposable and must be replaced after they have discharged. These include the zinc-carbon, zinc chloride, alkaline, mercuric oxide, silver oxide, lithium, and zinc-air types. Nicads and some of the disposable batteries can be represented by letters such as AAA, AA, C, sub-C, D, and N.

In some instances, such as in automobiles, motorcycles, boats, and planes, the selection of a battery as a power source is determined not only by one or more electric motors, but by the fact that the battery must also be able to handle a number of different kinds of loads, such as lights, communications equipment, heaters, and so on.

Lead-Acid Batteries

One of the most common and widely known of the entire family of batteries is the lead-acid, in use now for more than 140 years. In its basic form it consists of two lead plates immersed in a water solution of sulfuric acid. While both plates are lead they are chemically different. One of the plates is pure sponge lead and is the negative electrode. The other is lead peroxide and is the positive electrode. This cell develops about 2.15 V. For a lead-acid battery, six of the cells are connected in series, and the terminal voltage is 12.9 but is commonly referred to as a 12-V battery.

The 12-V lead-acid battery can be connected in series with another, supplying an output of 24 V. Pairs of 12-V batteries are wired in series and then housed in a single container with a 24-V output. It can also be connected in parallel, and if both batteries are identical, can double the current capability. Single lead-acid cells are available and can also be supplied as 2.5-Ah D-sized cells.

The life expectancy of a lead-acid battery is determined by the weakest cell in the series combination, and it is this cell that determines the current capability of the series group.

The charging rate of lead-acid batteries can be categorized as rapid, quick, standard, and trickle. For each of these the charging voltage, current, and charging time are listed in Table 2-1. In modern lead-acid cells the acid electrolyte is sealed in. To prevent splashing or accidental acid loss, the acid may be gelled; thus the battery can be held in any position.

Table 2-2 lists some characteristics of lead-acid batteries.

Nickel-Cadmium Cells

Like the lead-acid type, the nickel-cadmium, or nicad, is a storage unit, can be frequently recharged, and is a true secondary type. Unlike lead-acid batteries commonly identified by voltage output, nicads carry the same letter designations as their zinc-carbon counterparts and are interchangeable with them. Thus, nicads are available in sizes AAA, AA, C, sub-C, and D. They are also available in miniature sizes.

TABLE 2-1. CHARGING RATE OF LEAD-ACID BATTERIES

Type of charge	Charging voltage (volts DC per cell)	Charging current (% of cell capacity)	Charging time (hours)
Rapid	2.55–2.65	100	1–3
Quick	2.50–2.55	20–50	12–20
Standard	2.45–2.50	10–40	10–18
Trickle	2.28–2.32	10–20	Continuous

TABLE 2-2. CHARACTERISTICS OF LEAD-ACID BATTERIES

Memory	None
Current capacity	Rated in amperes
Cost	Moderate
Weight	1 pound or more
Connections	Parallel or series
Shelf life	50% power loss in less than 1 year

Battery Types and Their Characteristics

The output of a single cell, of which a cross-sectional view appears in Figure 2-9, is 1.25 to 1.4 V when measured under open circuit conditions with a high-resistance voltmeter.

The advantage of nicads is that for a given size they are lighter and can store and deliver more electrical energy than lead-acid types. Table 2-3 lists their charging rates.

The discharge rates of the nickel-cadmium and the lead-acid batteries follow a similiar curve, as indicated in Figure 2-10. Both battery types have a fairly flat discharge curve for the first 4 hr, after which they drop their output voltage under load sharply. Table 2-4 lists nicad characteristics.

Zinc-Carbon Cells

The zinc-carbon cell is one of the more widely used of the disposable types. Sometimes called a dry cell, it consists of two electrodes as shown in the cross-sectional view in Figure 2-11. The positive electrode is a carbon rod that extends from the top down through the center of the cell, almost to the bottom of a cylindrical container. The outer shell of this container, made of zinc, is the negative electrode. The terminals of the two electrodes are at the top and bottom of the unit. The electrolyte is a moist paste made of a mixture of granulated carbon and manganese dioxide.

Zinc-carbon cells are available in four sizes: D, C, AA, and the 9-V. The dimensions of each of these cells are supplied in Table 2-5.

The terminal voltage of these cells is 1.5, but the cells can be connected in series prior to being packaged in their container. Typical voltage values are 6.0, 9.0, and 15.0. Some of the larger units are high-voltage types and have outputs of 240 and 510 V. The electrical characteristics of this cell are listed in Table 2-6.

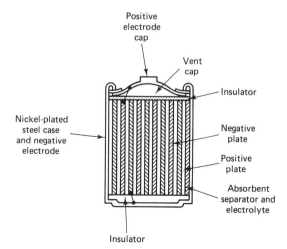

Figure 2-9. Exposed view of a nicad.

TABLE 2-3. CHARGING RATE OF NICKEL-CADMIUM BATTERIES

Type of charge	Charging voltage (volts per cell)	Charging current (% of cell capacity)	Charging time (hours)
Rapid	Depends on cell	Depends on cell	1–3
Quick	2.50–3.00	20	3–5
Standard	2.50–3.00	10	14
Trickle	2.50–3.00	5	Continuous

TABLE 2-4. CHARACTERISTICS OF NICADS

Basic type	Rechargeable
Current capacity	80 mAh (9-V battery)
	4 Ah (size D)
Shape	Cylindrical and rectangular
Compatibility	Interchangeable with comparable dry cells
Negative terminal	Case (nickel-plated steel)
Positive terminal	Circular metal plate on top
Electrodes	Positive electrode: nickel hydroxide
	Negative electrode: cadmium hydroxide
Electrolyte	Nonreplaceable potassium hydroxide (alkaline)
Venting	Two types: vented and hermetically sealed
Open-circuit voltage	1.25 V to 1.4 V
Load voltage	Essentially flat over current discharge range
Internal resistance	Very low, less than 100 mΩ
Internal capacitance	High
Series connection	Yes
Parallel connection	Not recommended
Recharging time	Typically 14 to 16 hours
Charging	Battery should be completely discharged before recharging

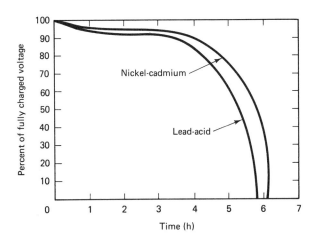

Figure 2-10. Operating characteristics of lead-acid and nicad batteries.

Battery Types and Their Characteristics

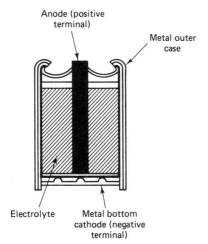

Figure 2-11. Cross-sectional view of zinc-carbon cell.

TABLE 2-5. DIMENSIONS OF ZINC-CARBON CELLS

Designation	D	C	AA	9 V
Voltage (V)	1.5	1.5	1.5	9.0
Diameter (in.)	1.344	1.0	0.560	—
Height (in.)	2.406	1.951	1.969	—
Length (in.)	—	—	—	1.031
Width (in.)	—	—	—	0.656

TABLE 2-6. ELECTRICAL CHARACTERISTICS OF ZINC-CARBON CELLS

Voltage	1.5 V
Positive electrode	Carbon
Negative electrode	Zinc
Rechargeability	Not rechargeable
Electrolyte	Granulated carbon and manganese dioxide
10-h current	60 mA (approx.)
100-h current	10 mA (approx.)

This battery can deliver approximately 60 mA for a 10-hr period, but for a lighter load it can supply 10 mA for 100-hr. However, the terminal voltage shows a sharp decline, as indicated in Figure 2-12. The voltage of the cell is capable of recovery if the load is intermittent, giving the cell a chance to recover. Table 2-7 lists the maximum current the cell can supply and some data about its internal resistance.

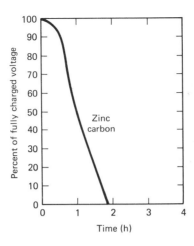

Figure 2-12. Zinc-carbon cell characteristic curve.

TABLE 2-7. MAXIMUM CURRENT AND RESISTANCE OF ZINC-CARBON CELLS

Cell size	Maximum current (amperes)	Internal resistance (ohms)
AA	4.6	0.31
C	5.4	0.28
D	6.6	0.23
9 V	0.5	19

Alkaline Cells

The alkaline cell is a disposable dry cell similar to the zinc-carbon. Its electrolyte consists of a solution of potassium hydroxide, and its positive terminal is made of zinc cylinders ending in a button-type connector on top of the cell. Its cathode uses cylinders of manganese dioxide, which terminate in the bottom of the cell and form the negative terminal.

The alkaline cell is superior to the zinc-carbon in a number of respects. It is capable of delivering more current; it has a longer current-storage capability; and it can perform better at lower and higher temperatures. Construction details of this cell are shown in Figure 2-13. It finds application in motor driven toys and portable tape recorder/players. Table 2-8 lists the characteristics of this cell, and Table 2-9 supplies its dimensions.

Mercury Cells

The mercury cell, a secondary type, is desirable for its small size. Housed in a steel case, it is noted for its dependability. Various types are shown in cross-section in Figure 2-14.

Battery Types and Their Characteristics

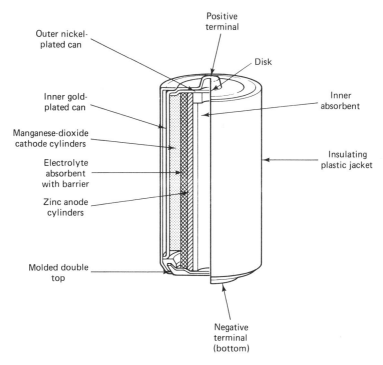

Figure 2-13. Details of alkaline cell.

TABLE 2-8. CHARACTERISTICS OF ALKALINE CELLS

Weight	About 50% more than a comparable zinc-carbon cell
Voltage	1.5 V (except 9-V transistor type)
Negative terminal	Manganese dioxide
Positive terminal	Powdered zinc
Electrolyte	Potassium hydroxide gel
Rechargeability	Units are primary cells and are not rechargeable
Internal resistance	Very low
Impedance	Very low, less than carbon-zinc
Shape	Available as cylinder or button type

TABLE 2-9. DIMENSIONS OF ALKALINE CELLS

Size	Voltage (volts)	Diameter (inches)	Height (inches)	mAh
D	1.5	1.32	2.39	14,000
C	1.5	1.0	1.951	7,000
AA	1.5	0.56	1.969	2,100
AAA	1.5	0.410	1.75	1,000
		Width	Length	
9-V transistor		0.65	1.031	1.906	550

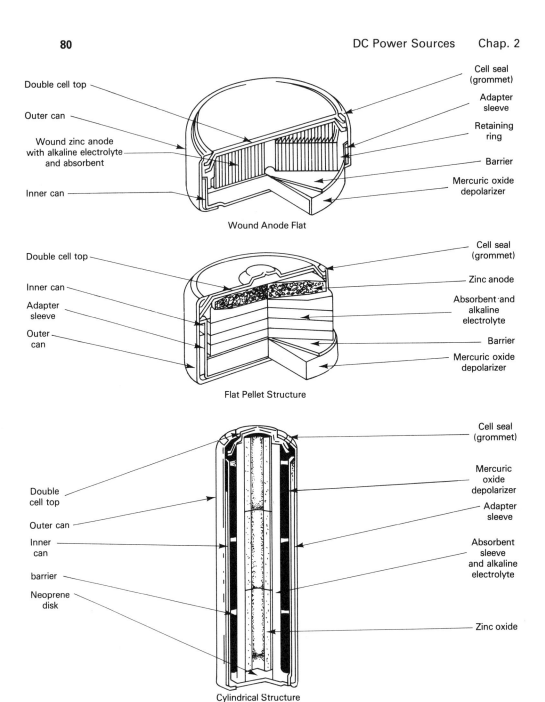

Figure 2-14. Various types of mercury cells.

The electrolyte, as in the case of alkaline cells, is potassium hydroxide. The negative plate is made of mercuric oxide, from which the cell derives its name. The positive plate is amalgamated zinc. The voltage of a single cell is 1.35 V. Table 2-10 lists the characteristics of mercury cells.

The size of the mercury cell is such that its diameter and height dimensions, in inches, are all in fractional decimals. In terms of milliampere-hours the current capacity extends from a low of 80 mAh to as much as 1,000 mAh. A pair of single cells can have a wide disparity in current capacity. This characteristic and its physical dimensions are detailed in Table 2-11.

The cells can be connected in series and are packaged to supply 5.4, 6.75, 8.4, 9.8, and 12.6 V, and other voltages as well. The advantage of the mercury cell, aside from being extremely compact, is that it has a longer shelf life than the carbon types. Mercury cells aren't designated by voltage or size but by manufacturers' part numbers.

Lithium Cells

Ordinarily, a cell is made with two different chemicals, one for its positive plate, the other for the negative. The lithium cell, however, uses lithium for one, but several different kinds for the other, as indicated in Table 2-12.

TABLE 2-10. CHARACTERISTICS OF MERCURY CELLS

Voltage	1.4 V, open circuit, slightly lower than carbon-zinc or alkaline; voltage does not decrease until cell is discharged
Rechargeability	Not rechargeable; mercury cells are primary types
Positive electrode	Red mercuric oxide
Negative electrode	Zinc-mercury
Electrolyte	Solution of potassium hydroxide and zinc oxide
Shelf life	Better than zinc-carbon
Short-circuit current (AA cell)	10.3 A
9-V cell	1.7 A

TABLE 2-11. SIZES OF MERCURY CELLS

Voltage (volts)	Diameter (inches)	Height (inches)	mAh
2.7	0.665	0.607	250
5.6	0.600	0.787	110
5.6	0.500	0.787	150
1.35	0.625	0.440	500
1.35	0.455	0.135	80
1.35	0.625	0.645	1,000
5.4	0.675	1.767	500
1.35	0.455	0.210	190
1.35	0.615	0.238	250
4.05	0.660	0.845	250

TABLE 2-12. CHEMICAL COMPONENTS FOR LITHIUM CELLS

Cell chemistry	Open-circuit voltage (volts)
Lithium–thionyl chloride	3.7
Lithium–vanadium pentoxide	3.4
Lithium–silver chromate	3.3
Lithium–manganese dioxide	3.0
Lithium–sulfur dioxide	2.9
Lithium–carbon monofluoride	2.8
Lithium–iodine	2.8
Lithium–lead-copper sulfide	2.2
Lithium–copper sulfide	2.1
Lithium–iron sulfide	1.8
Lithium–copper oxide	1.8

Although the lithium cell isn't well known, it does have a number of advantages. It has a shelf life of more than 10 years and is capable of producing substantial electrical energy for its size. Furthermore, it isn't temperature-sensitive. The long shelf life can make up for the fact that the lithium cell isn't rechargeable. It is used in cameras that have a motor function. The various characteristics of this cell are listed in Table 2-13.

These cells have a current capacity ranging from 60 to 250 mAh. They are button types having fractional decimal dimensions as indicated in Table 2-14. The maximum diameter and height, as indicated in this table, is each less than 1 in. The current output is small and is measured in milliamperes.

Silver Oxide Cells

The full name for this cell is silver oxide–alkaline zinc. It is available in two different forms, depending on the electrolyte used. One of these is sodium hydroxide, the other potassium hydroxide. Both are basic, as opposed to acidic, as in the lead-acid cell. The sodium hydroxide–type cells are characterized by a long working life, that is, from 2- to 3-years.

TABLE 2-13. CHARACTERISTICS OF LITHIUM CELLS

Voltage discharge	Flat
Rechargeability	Does not recharge
Internal resistance	Very low
Impedance	Low
Cathode (negative element)	Manganese dioxide and carbon black
Anode (positive element)	Lithium foil
Electrolyte	Lithium perchlorate in propylene carbonate
Reaction to temperature	Little or none
Voltage per cell	3.0 V
Type	Dry cell button

Battery Types and Their Characteristics

TABLE 2-14. DIMENSIONS OF LITHIUM CELLS

Voltage (volts)	Diameter (inches)	Height (inches)	mAh
3.0	0.460	0.420	160
6.0	0.510	0.990	160
3.0	0.630	0.079	60
3.0	0.787	0.063	75
3.0	0.787	0.098	145
3.0	0.787	0.126	190
3.0	0.965	0.079	120
3.0	0.965	0.118	250

The cell is rated at 1.5 V and is quite small, with dimensions in decimal fractions of an inch, as shown in Table 2-15.

Zinc Chloride Cells

This cell is a variation of the zinc-carbon type described earlier and may be considered an improvement on them, in that they have an increased current capability on the order of 50%. One of its characteristics is that the cell voltage decreases during the time it supplies current under load. Sometimes, to distinguish them from zinc-carbon cells, they are known as heavy-duty types. As far as current capacity is concerned, they are rated below alkaline cells.

TABLE 2-15. DIMENSIONS FOR SILVER OXIDE CELLS (1.5 V THROUGHOUT)

Diameter (inches)	Height (inches)	mAh
0.455	0.110	70
0.455	0.122	70
0.455	0.081	35
0.305	0.143	38
0.305	0.210	70
0.374	0.141	45
0.375	0.102	67
0.311	0.102	24
0.311	0.102	24
0.455	0.135	83
0.455	0.137	90
0.610	0.190	250
0.455	0.210	180
0.311	0.082	16
0.267	0.084	15
0.311	0.141	38
0.455	0.165	120

The zinc-chloride cell has an open-circuit voltage of 1.5. The zinc can is the positive terminal, and the electrolyte, made of zinc chloride, has a negative polarity. The cell is a primary type, and theoretically isn't rechargeable. However, it can be partially, though not wholly, recharged.

Zinc-Air Cells

Another type of cell is the zinc-air, but the zinc is not one of the electrodes. Oddly, the negative terminal is air, with access to this gas gained by a tiny hole that is kept covered until the cell is ready for use. The positive terminal is the can, made of nickel-plated steel. The voltage measured across its electrodes is between 1.4 and 1.45, but when supplying current this potential drops to 1.3 V or possibly as low as 1.1 V. When supplying a load, the final voltage supplied remains fairly constant.

INTERNAL RESISTANCE OF A BATTERY

For circuits external to a battery, an electric current will flow from its minus terminal, through a load such as a motor, and then back to the positive terminal. No current is ever lost, and the same amount of current that leaves the battery always returns to it.

Inside the battery the current flows from the positive to the negative terminal, and in following this path the current passes through the resistance that exists between its electrodes. Because both I and R are present, there is a power loss inside the battery at the time it supplies current to a motor. In terms of a formula, the power loss is equal to

$$P = I^2 \times R$$

The internal resistance of a battery cannot be measured directly with a test instrument such as an ohmmeter, but it can be calculated indirectly. The internal resistance can be determined by regarding the battery as a DC generator, shown as E_g in Figure 2-15, and its internal resistance considered equivalent to an external resistor, shown as R_g. The load, R_L, is shunted across this series combination.

The same current, I, flows through these two "resistors," R_g and R_L, and so both will have an IR drop, a voltage. The sum of these two voltages will be equal to the battery voltage and can be expressed as

$$E_g = IR_g + IR_L$$

When a battery is used for driving a motor, its internal resistance increases, producing an increased voltage drop across that resistance. The terminal voltage of the battery decreases by the same amount. There is also a power loss developed by this internal resistance, and in time the power loss increases, varying in amount from one battery type to another.

Battery Efficiency

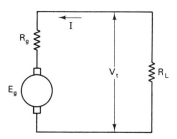

Figure 2-15. Internal resistance of a voltage source.

BATTERY EFFICIENCY

A battery can supply power to a load, such as a motor, and also develops a power loss across its internal resistance. The ratio of these two is the efficiency of the battery.

$$\eta = \frac{P_o}{P_i}$$

where

η = the efficiency,

P_o = the output power, and

P_i = the power developed in the battery.

The output power is equal to the input power minus the power loss caused by the internal resistance.

The efficiency of a battery (η) is maximum when the battery is new or freshly charged, but gradually decreases as the battery is used, or while it is allowed to remain unused for some time. The loss of power capability of a battery is referred to as its shelf life and is also a variable, depending on the battery type and the quality of its construction. Since the output power is always less than the available battery power, the efficiency always has a value of less than 1 or 100%. Efficiency can be expressed as a decimal, or, multiplied by 100, as a percentage.

By transposing the terms of the formula for efficiency, two new formulas can be developed. The first of these is

$$P_o = \eta \times P_i$$

This formula indicates that the available output power is equal to the total power capability of the battery multiplied by its efficiency.

Another variation of the efficiency formula is to write it as

$$P_i = \frac{P_o}{\eta}$$

This formula shows that the total power of the battery, assuming no power losses, is equal to the output power divided by the efficiency.

ELECTRONIC POWER SUPPLIES

An electronic power supply is a component connected to the AC power line either directly or via a power transformer. The AC is rectified into pulsating DC, which can then be used to drive a motor. In some cases the pulsating DC is fed into a filter whose output is a smoother form of DC. The transformer is either a step-up, a step-down, or a one-to-one type.

Electronic power supplies can work as chargers for lead-acid or nicad batteries and are available commercially. In some instances the charger is equipped with a meter to indicate how much current is flowing into the battery. Figure 2-16 shows a simple circuit for charging these two battery types. The primary winding of the transformer is connected to the AC power line, the typical voltage of which for in-home use is 120 V. The primary winding may be fused, and in some motor power supplies the secondary is fused as well.

The secondary of the transformer is an AC voltage, which is rectified by a diode semiconductor, D. The resistor, R, following the diode is used to limit the amount of current flow into the battery and can be either a fixed or variable unit. The problem with using resistor R is that the entire charging current flows through it; thus it develops a power loss in the form of heat.

The purpose of resistor R is to adjust the voltage so the charging potential across a battery is a little larger than its terminal voltage.

Charge resistor R can be eliminated by using a transformer having one or more taps on the secondary winding (see Figure 2-17). As the connection to the tap is moved down toward the bottom of the transformer, the amount of power-supply output voltage increases. If the transformer has a number of taps, the output voltage can be fairly well controlled without the need for a power loss. The circuit in Figure 2-17 also shows a capacitor working as a filter connected across the DC output. The capacitor helps supply a smoother DC output.

Assuming the batteries are discharged, the maximum current flows when the batteries are first put on charge. As the batteries charge they build up a voltage

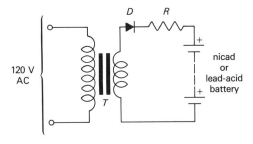

Figure 2-16. Charger suitable for either lead-acid or nicad batteries.

Figure 2-17. Half-wave rectifier using tapped transformer. (Courtesy RCA)

that is opposite in polarity to the DC voltage supplied by the charger, and so the charging current is gradually reduced.

Dual-Polarity Power Supplies

With some motors reversal of rotation of the armature can be obtained by transposing the polarity of the voltage supply. One way of doing this is with the circuit in Figure 2-18, consisting of a pair of diodes identified here as A and B. Current flow through the diodes depends on the polarity of the input AC waveform. Without a switch the output voltage is alternately plus or minus for equal amounts of time. Separate switches can be inserted in the output lines of the diodes to obtain manual control of armature rotation.

Determination of the Charge Rate

To determine the correct charging rate, multiply the ampere-hour capacity of the cell by 0.1. If the battery is rated at 800 mAh, then 800 × 0.1 = 80 mA. The easiest way to determine the amount of charging current is to use a DC milliammeter. This can be inserted anywhere in the circuit connected to the secondary of the transformer. In the case of lead-acid batteries, the charging current is usually in amperes.

Even if a battery is not used, it will need to be recharged. For nicads, the current capability is reduced by 50% in about three months; the same amount of current capability is lost in about eight months or more by the lead-acid type.

The disadvantage of the single-diode half-wave rectifier is that for 50% of the time no current is being delivered by the charger to the battery, as indicated in Figure 2-19. The frequency of the DC output pulses is the same as the frequency of the AC input to the power transformer. The output of a battery is always pure DC, while that of an electronic power supply, without a filter, is fluctuating or pulsating DC. The output of an electronic power supply can be improved so that it is also pure DC, but this requires additional circuitry, including a filter, and possibly a voltage regulator as well. Whether this is necessary depends on the motor to be used and the sensitivity of its load to voltage variations.

A more efficient type is the full-wave rectifier in Figure 2-20. The frequency of the DC pulse output is twice that of the frequency of the AC input (Figure 2-21). The current flow from the charger to the battery or the motor is still in the form of pulses, but these are continuous with no time interval between them.

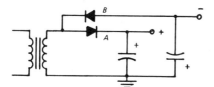

Figure 2-18. Half-wave power supply with plus and minus voltage outputs. (Courtesy RCA)

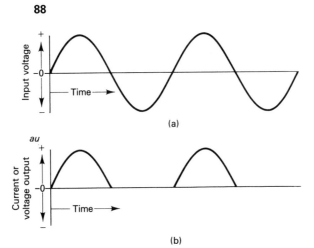

Figure 2-19. Waveform of AC input to half-wave rectifier (a) and output waveform (b).

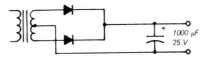

Figure 2-20. Full-wave supply with low-voltage output for motor use. (Courtesy RCA)

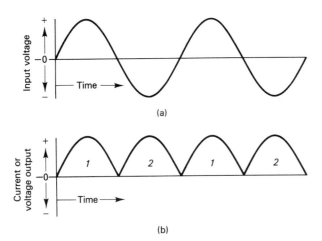

Figure 2-21. Input (a) and output (b) of full-wave rectifier.

The same circuits, the half-wave rectifier and the full-wave, can also be used to deliver the operating power for motors. The power supply must be capable of supplying the needed amount of DC voltage and operating current. The voltage must be correct, but the operating current can be that required by the motor, or higher. If it is less, the motor will either not operate or may not be able to supply enough torque.

Voltage Regulators

A filter capacitor can also be connected across the battery. This has the effect of making the charging current a smoother DC, but it is not always included with chargers. It also has the effect of degrading the voltage regulation.

Voltage Regulation

Voltage regulation is always a fractional decimal but becomes a percentage through multiplication by 100, as indicated in the following formula.

$$\% \text{ Voltage Regulation} = \frac{E_{\text{no load}} - E_{\text{full load}}}{E_{\text{full load}}} \times 100$$

Example:

A half-wave power supply has a no-load output of 24 V. When connected to a motor it drops to 21 V. What is the percentage regulation of this supply?

Solution:

$$\% \text{ Voltage Regulation} = \frac{E_{\text{no load}} - E_{\text{full load}}}{E_{\text{full load}}}$$

$$= \frac{24 - 21}{24} = \frac{3}{24} = 0.125 = 12.5\%$$

The lower the percentage, the better the voltage regulation.

VOLTAGE REGULATORS

The output voltage of an electronic power supply and its regulation depends on a number of factors: the type of filter used in the supply, whether the motor load is constant or variable, and the current-passing ability of the rectifier. The rectifier is a current-handling device, and the larger its resistance the greater the voltage drop across it. This voltage drop takes precedence over the output voltage, and its amount is subtracted from the output.

A power supply having perfect regulation will have constant output voltage regardless of variations of the load on the motor. There are various electronic circuits that can be used to improve regulation, including the Zener diode regulator, the forward-biased diode regulator, regulators using transistors, and silicon-controlled rectifier (SCR) regulators.

While electronic regulators could be used with batteries, batteries in turn can have improved regulation through the simple expedient of recharging. However, with use and in time a battery may not accept a full or partial charge, and its regulation will become poorer. With an electronic power supply there is no dependency on charging, and a voltage regulator circuit will help maintain fairly good regulation with motor load variations.

SURGE LIMITING

A motor imposes the highest load when first connected to a voltage source. It takes a small but finite time for the motor to develop a counter EMF, (counter voltage) but until it does so there is a large surge of current. This current flow can exceed the surge-current passing capability of the diode rectifier. To limit the current to a safe value, a small ohmage resistor is put in series with the diode.

At the time the motor is first turned on, not only is there no existing counter EMF, but the input filter capacitor is uncharged and so contributes to the current demand. The power supplies in Figures 2-18 and 2-20 are transformer types and with these the surge limiter is often omitted, since the secondary winding of the transformer is assumed (because of its wire resistance) to act as a limiter. However, the surge limiter is sometimes included as a precaution.

ZENER-DIODE REGULATOR

Figure 2-22 illustrates the circuit of a simple Zener-diode regulator. With this arrangement the input is unregulated DC, possibly that supplied by a half-wave power supply using a capacitor filter.

The operating behavior of the Zener diode is illustrated in the graph of Figure 2-23. This graph shows that the voltage across the Zener is flat, that is, remains

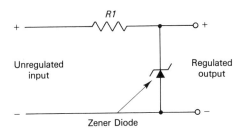

Figure 2-22. Zener voltage-regulating circuit. (Courtesy RCA)

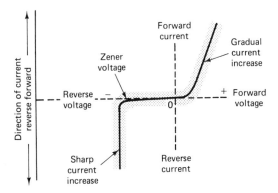

Figure 2-23. Current characteristics of Zener diode.

Zener-Diode Regulator

unchanged from zero to the left with the application of a reverse bias. A point will be reached when the application of further reverse bias will result in a sharp increase in current. The voltage point, known as the Zener voltage, can be as little as 3 or as much as 150 V, depending on the way the Zener is made. Its power rating is from about 200 mW to more than 50 W.

A Zener can also be used to regulate the input voltage when it is AC, as shown in Figure 2-24. Here a pair of Zeners are placed back-to-back so both halves of the AC input-voltage waveform are maintained at a steady level. Though the frequency of the AC power line remains remarkably constant, the voltage can vary depending on the loads imposed on that line.

Series Zeners

The required voltage input to a motor will determine the type of Zener to use. Like resistors, capacitors, or other electronic components, Zeners can be connected in series, as indicated in Figure 2-25. Here a pair of 33-V Zeners are wired in series, capable of supplying regulation for a total input of 66 V. If an 18-V Zener is included in the series hookup, a total of 84 V can be regulated. The Zeners can be connected in any order. Thus, the 18-V unit can be placed between the two 33-V diodes.

There are some restrictions for Zener connections. The amount of current flowing through these diodes is limited by the maximum current rating of any one

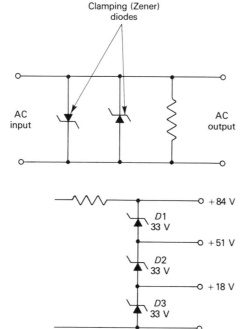

Figure 2-24. Zeners used for regulating input voltage.

Figure 2-25. Stacked Zener diodes supplying multiple output voltages (Courtesy RCA)

diode. Thus, in the series arrangement of Figure 2-25, there are three possible voltage taps: 18 V, 51 V (18 + 33 = 51), and 84 V (18 + 33 + 33).

Cascaded Zener regulation. The simple Zener circuit shown earlier in Figure 2-22 can be improved by the cascade arrangement of Figure 2-26. Here the regulated 12-V output of the first Zener is further regulated by the second Zener.

Low-voltage motor control. Regulating low voltages can be done with the help of the circuit shown in Figure 2-27. A pair of Zeners are used so that regulation can be obtained within narrow limits. Thus the first Zener regulates at 6.8 V, the second at 5.6 V. The motor, designated as the load in this case, is connected between the two diodes as indicated. The regulated voltage is the difference between these two, or 6.8 − 5.6 = 1.2 V. By careful selection of the Zeners the regulated voltage can be even smaller than this. The same circuit can also be used for larger voltages. Zeners work well in series; however, a parallel connection isn't recommended.

Diode Voltage Control

A solid-state diode can be used instead of a Zener for low-voltage control, as shown in Figure 2-28. This circuit shows a fractional decimal voltage of 0.7 and is suitable for voltage regulation between 0.5 and 1.1 V. The diode in this case supplies voltage regulation that is not as good as that achieved with Zener diodes.

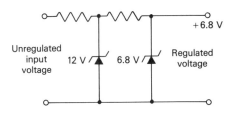

Figure 2-26. Cascaded Zener-diode regulators. (Courtesy RCA)

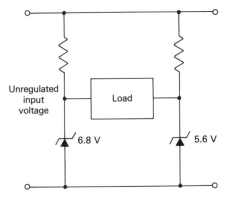

Figure 2-27. Low-voltage motor control. (Courtesy RCA)

Zener-Diode Regulator

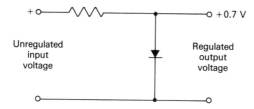

Figure 2-28. Regulating very low voltage by means of a forward-biased diode. (Courtesy RCA)

Transistor Voltage Regulators

Power-supply voltage regulation for use with motors can be obtained by a feedback circuit. This circuit senses any change in output voltage and feeds back a signal that cancels any variation. While there is a momentary increase or decrease in the source voltage the fluctuation is so quick that motor operation is not affected.

The basic arrangement consists of a reference voltage, a voltage required by the motor. The output of the power supply is compared with this reference. If the DC output voltage and that of the reference are the same, no feedback signal is used. If there is a voltage difference, that is, if the source voltage is greater or less than the reference, the difference between the two is fed back to the source as a correction, causing the source voltage to increase or decrease as required.

An arrangement of this kind is shown in Figure 2-29. The reference voltage is fed into a circuit referred to as a difference amplifier, a combination error detector and DC amplifier. A portion of the source voltage is taken from a tap point of two series resistors connected across the regulated source voltage. This voltage is compared in the difference amplifier with that received from the reference voltage. Any difference voltage is added to or subtracted from the source voltage. As soon

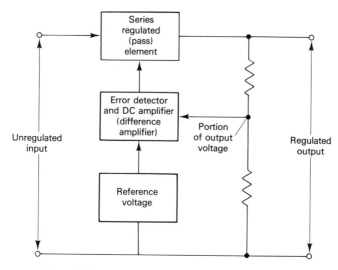

Figure 2-29. Block diagram of series voltage regulator.

The Series Zener and Transistor Regulator

Figure 2-30 shows the details of a series voltage regulator. The transistor, Q1, is connected in series with the motor, using the negative leg. The collector of the transistor is connected to the negative terminal of the unregulated source voltage, the emitter to the negative terminal of the regulated output. Because of this arrangement the motor current flows through the transistor, which must be selected to accommodate the current requirements of the motor.

The emitter-follower transistor stage works like a variable resistor controlled by the Zener. Although just a single transistor is shown here, it may actually consist of several. The problem with this circuit is that the output resistance varies slightly with the load. Thus, there is a moderate difference in the output voltage depending on whether the motor is heavily or lightly loaded, with the voltage decreasing for a load that causes the motor current to rise.

Regulation With a Differential Amplifier

The circuit of Figure 2-31(a) is that same as that previously shown in Figure 2-30, except that a differential amplifier is now included. As illustrated earlier, the transistor is in series with the negative leg of the power supply and works as an emitter follower (common emitter). A variation of this circuit is shown in Figure 2-31(b).

The differential amplifier in both circuits works as a difference or error amplifier. This amplifier has two inputs. One of these is the regulated voltage taken from the tap between the Zener and its connecting resistor. The other input is from the series resistors connected across the output. The differential amplifier senses the difference between these two voltages and uses that difference voltage to control the transistor.

Shunt Transistor Voltage Control

For very small currents, a transistor having a limited current capability can be used. For higher currents, the limited-current transistor can be replaced by a power transistor. In any event, the transistor must be capable of passing the maximum

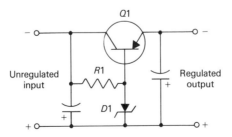

Figure 2-30. Circuit of series voltage regulator. (Courtesy RCA)

Zener-Diode Regulator

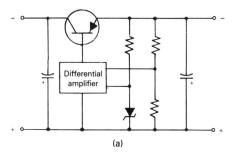

(a)

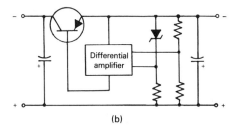

(b)

Figure 2-31. Circuits a and b show two possible locations of the differential amplifier. (Courtesy RCA)

current demand of the motor. If a single power transistor is inadequate, a number of them can be wired in shunt, as in Figure 2-32. However, for large currents the differential amplifier may not be able to supply adequate control.

Darlington Motor Control

In some motor control circuits the amount of motor current may be equal to or in excess of the current capability of the shunt-connected power transistors. In that case the shunt power transistors can be replaced by a Darlington amplifier, as illustrated in Figure 2-33. In this circuit, transistor Q2 works as an emitter follower,

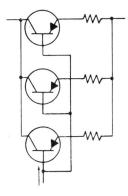

Figure 2-32. Transistors in shunt for increased current-handling capacity. (Courtesy RCA)

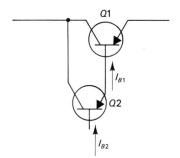

Figure 2-33. Darlington configuration for reducing motor current-control requirements. (Courtesy RCA)

while transistor Q1 is the series-regulating transistor. With this circuit the amount of control current is greatly reduced.

The Shunt Regulator

One method of eliminating the current limitation of the series transistor is to use a shunt regulator, whose block diagram and schematic diagram are shown in Figure 2-34(a) and (b). These circuits are simpler than the series type but are quite inefficient. However, they do provide overload and short-circuit protection.

In the block diagram in Figure 2-34(a) the output voltage remains constant, since the shunt element bypasses more or less current, depending on motor load variations or changes in the input source voltage. When such changes occur, the voltage drop across resistor R1 varies accordingly.

The shunt element is wired in parallel with the motor input. The shunting circuit, as in the case of the series regulator, may consist of two or more transistors. It may also include a Darlington circuit, as indicated in Figure 2-34(b).

MOTOR GENERATORS

A motor generator is an electromechanical device used in industry for producing either an AC or DC voltage output. It is needed where a motor requires a substantial DC input not capable of being supplied by batteries or an electronic power supply. It is also used where an AC output is required but there is no access to a power line, as in the case of a ship.

There are four possible motor generator arrangements. These include

1. DC motor driving an AC generator
2. DC motor driving a DC generator
3. AC motor driving an AC generator
4. AC motor driving a DC generator

In the case of item 2 listed above, this combination is needed when the available DC voltage is inadequate and must be increased so as to be able to drive

Motor Generators

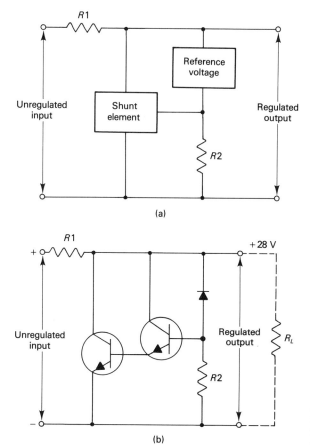

Figure 2-34. Basic shunt regulator: (a) Block diagram; (b) schematic. (Courtesy RCA)

a certain type of motor. Unlike AC, DC voltage cannot be stepped up through the use of a transformer. The drawing in Figure 2-35 shows the basic arrangement of a motor generator.

The drive shaft of the motor may be mechanically linked to a generator by a coupling device or through a gear train. In the case of the arrangement indicated in the drawing, the armature of the motor and that of the generator will rotate at the same speed. A gear train can be used instead so that the generator armature rotates at a higher or lower speed.

Advantages and Disadvantages of the Motor Generator

The motor generator supplies simple output-voltage control. As a source voltage, it has excellent regulation. It does not supply the pure DC output obtained from batteries, but it is better in this respect than power supplies not equipped with filters or voltage regulating circuits. It has a much higher ripple frequency than the output of a half-wave or full-wave rectifier type power supply. Because of this,

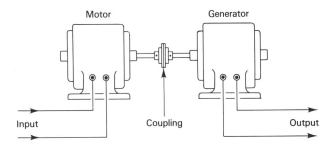

Figure 2-35. Motor generator.

frequency filtering is simple. The motor generator does not need rectifiers and can be operated with either a DC or AC input. It can also be used where batteries would be impractical and no AC source is available.

The motor generator is large and heavy, and is usually more expensive than an equivalent combination of batteries or an electronic power supply. It is noisy and can cause interference to radio and television receivers. Repairing the motor generator can be difficult, and it requires lubrication and replacement of brushes.

THE DYNAMOTOR

As a source voltage, the dynamotor has excellent regulation, but it does not supply the pure DC output obtained from batteries. The dynamotor could be called a DC-to-DC machine, for its input is a low-voltage DC such as that supplied by a lead-acid storage battery. Its generator output is high-voltage DC. Since non-varying DC cannot be used with a power transformer, the dynamotor works as a substitute.

The dynamotor, Figure 2-36, shows that the armature and field coils are common to both units, the motor and the generator. Although the illustration shows a single armature, there are two independent windings with each occupying slots in the armature core.

The unit is equipped with two commutators: one for the DC input, the other for the generator. The motor commutator permits the battery to supply current to the rotating armature. The generator commutator works as a rectifier, converting the AC supplied by the armature to DC. In both instances, for the motor and generator commutators, input and output currents are delivered and supplied to a load via brushes.

The DC-to-DC dynamotor is used in industry where high voltage and high direct current are required. There are also AC dynamotors, which have a DC input and an AC output. The input to the motor section is the same as in the DC-to-DC dynamotor. However, the AC output does not have a commutator. Instead, the alternating current supplied by the generator section is brought out to a pair of slip rings. These also use brushes for the delivery of the alternating current to an AC motor. The DC-to-AC dynamotor is used in transportation where no source of AC is available.

The Rotary Converter

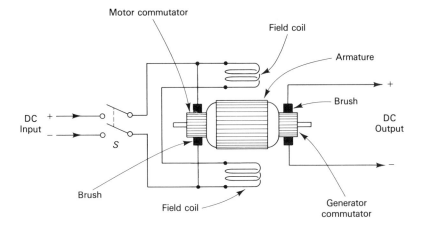

Figure 2-36. Dynamotor.

THE ROTARY CONVERTER

The rotary converter, shown in Figure 2-37, resembles the dynamotor as far as the field coils and commutator for the DC input are concerned. The output is AC and is obtained via a pair of slip rings. Unlike a dynamotor the rotary converter has just a single armature, which works for both the generator and motor sections of the unit.

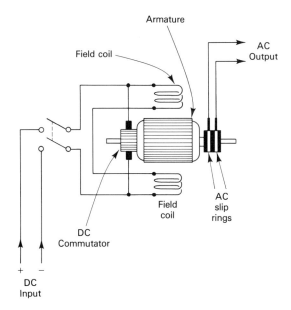

Figure 2-37. Rotary converter.

Because of its construction, the function of the converter can be transposed, that is, an AC source voltage can be applied to the slip-ring output and DC can be obtained from the commutator.

Since the input and output are so closely related, the output voltage can fluctuate with voltage changes at the input. When it works as an AC motor it is of the synchronous type, and in this mode is sometimes referred to as a synchronous converter.

chapter three

AC for Motors

The waveforms in Figure 3-1(a) are those of DC voltages, while those in (b) represent AC. The waveforms are identical, with the only difference being the location of the zero reference line. The voltages do not go below this line in the first drawing, but do so in the second. This indicates that DC voltages do not change polarity and that the currents associated with these voltages do not alter their direction of movement. In the AC waveforms the voltages change polarity, and the resulting currents alternately change the direction of their movement in step with the changes in voltage polarity.

WAVEFORM VARIATION

The sine wave in Figure 3-2 is just one of many different kinds of waveforms, and although it is one of the most common types used for AC motors, it can be DC as well as AC. Other waveforms are illustrated in Figure 3-3. That in Figure 3-3(a) is known as a sawtooth because of its fancied resemblance to the teeth of a saw. The square wave in Figure 3-3(b) reaches its maximum extremely rapidly, theoretically in zero time. It maintains this peak value for a selected period of time and then drops to the zero reference line, also extremely rapidly. A voltage of this kind is referred to as a *pulse waveform* and is sometimes used to step a motor armature through a preselected number of degrees.

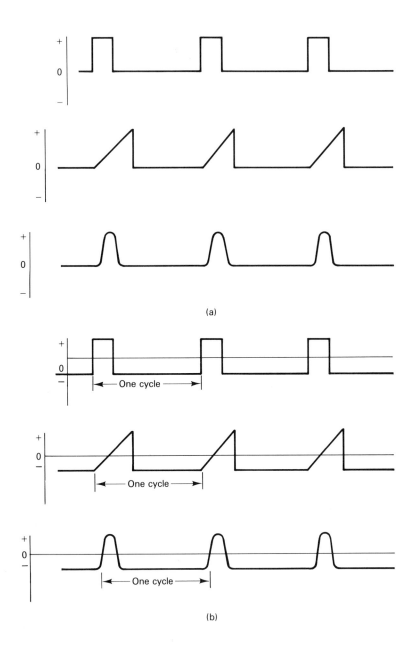

Figure 3-1. DC Voltages (a); AC voltages (b).

The AC Sine Wave

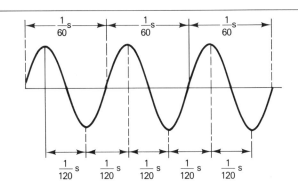

Figure 3-2. Graph of a 60-Hz sine wave.

(a)

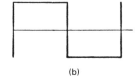
(b)

Figure 3-3. AC waveforms can have different shapes: (a) sawtooth; (b) rectangular.

VOLTAGE AND CURRENT MEASUREMENTS

The measurement of a pure DC voltage (pure in the sense that it does not vary and remains constant) carries no complications, since such a voltage is at its peak, remaining at its peak until the voltage is turned off. Both DC and AC need a fixed reference, with any voltage above that reference considered positive, or plus; voltages below it are referred to as negative, or minus. The difficulty with making measurements of an AC voltage is that it isn't fixed but is a variable amount. Both DC and AC voltages are usually measured with respect to a fixed reference, and quite often this is a zero voltage or a ground that is at zero voltage. A voltage, whether positive or negative, can also be measured with respect to some other voltage, also either positive or negative.

THE AC SINE WAVE

For AC motors the shape of its source voltage is commonly a sine wave, measured in angular degrees (deg.). A single cycle of a sine waveform, regardless of its frequency, lasts for 360°, after which the wave repeats. During a single cycle there are two peaks, one positive and the other negative; that is, there is one peak occurring at each 180°. But while peak voltage measurements are useful, they are not truly representative. In some instances a measurement is made from the positive peak to the negative peak and is referred to as *peak-to-peak,* or "p-p."

Average Values

Figure 3-4 depicts a sine wave with a number of vertical lines drawn in the positive half of the wave. By measuring the heights of these lines, known as ordinates, and averaging them, the result will be about 60% of the peak, or maximum, value of the wave. Using more lines to provide a more accurate sampling would show the average value to be 63.7% of the peak. Written in formula form

$$E_{average} = 0.637 \times E_{peak}$$

or

$$I_{average} = 0.637 \times I_{peak}$$

The illustration does not show actual values of voltage or current, nor is this required, since the peak can be arbitrarily considered as 100%, or 1. Because the formula is an $A = B \times C$ type, both sides can be divided by 0.637, supplying

$$E_{peak} = \frac{E_{average}}{0.637}$$

By assigning a value of 1 to the average voltage it becomes possible to establish a direct relationship between the peak voltage and the average voltage. The peak voltage, then, is 1.57 times the average voltage.

$$E_{peak} = \frac{1.0(E_{average})}{0.637} = 1.5698 = 1.57\, E_{average}$$

and

$$I_{peak} = 1.57 \times I_{average}$$

Note that

$$E_{peak} = 0.5\, E_{peak\text{-}to\text{-}peak}$$

or

$$E_{peak\text{-}to\text{-}peak} = 2 \times E_{peak}$$

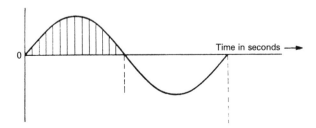

Figure 3-4. Method for finding the average value of a sine wave.

The AC Sine Wave

It is also possible to find the average value, given the peak-to-peak value, by

$$E_{average} = 0.3185 \times E_{peak\text{-}to\text{-}peak}$$

$$(0.637/2 = 0.3185)$$

$$I_{average} = (0.637 \times 2)(I_{peak} \times 2)$$

$$I_{peak} \times 2 = I_{peak\text{-}to\text{-}peak}$$

$$I_{average} = 1.274 \times I_{peak\text{-}to\text{-}peak}$$

Instantaneous Values of Sine Voltages and Currents

The instantaneous values of a sine wave of voltage or current are those that exist at a selected instant of time. Mathematically these are represented by the letters e (for voltage) or i (for current). It is possible to have as many instantaneous values as required. The value of an instantaneous voltage can be calculated by

$$e = E_{peak} \sin \omega t$$

where

t = the time in seconds, and

ω (the lowercase Greek letter omega) = $2\pi f$, in which

π = 3.14, and

f = the frequency in hertz (Hz) formerly cycles per second (cps).

Table 3-1 supplies the sine of angles from 0 to 360 deg. in 15-deg. steps.

TABLE 3-1. SINE VALUES OF ANGLES IN STEPS OF 15 DEG.

Angle (degrees)	Sine value	Angle (degrees)	Sine value
0	0.000	195	-0.259
15	0.259	210	-0.500
30	0.500	225	-0.707
45	0.707	240	-0.866
60	0.866	255	-0.966
75	0.966	270	-1.000
90	1.000	285	-0.966
105	0.966	300	-0.866
120	0.866	315	-0.707
135	0.707	330	-0.500
150	0.550	345	-0.259
165	0.259	360	0.000
180	0.000		

The sine of an angle, sometimes written as "sin" but always pronounced as *sine*, increases as the angle approaches 90°. Values of the sine above 180° simply repeat values lower than 180°. Thus, the sine of 195° is the same as the sine of 180° + 15°. Table 3-1 shows that the sine of 180° is 0 and that of 15° is 0.259. This is also the value of the sine of 195°.

The base or zero reference line in Figure 3-5 is shown in angular degrees. It is possible to select any angle along this base line and represent it by the Greek lowercase letter α. Then,

$$\alpha = \omega t$$

By substituting this in the formula for instantaneous values

$$e = E_{peak} \sin \alpha$$

Figure 3-6 shows two instantaneous values of voltage, $e1$ and $e2$, selected at random. For a particular sine wave, peak values and average values are fixed. Instantaneous values are fixed only in relationship to a particular instant of time. The same formula can be used for instantaneous values of current. It is essential to remember that the arithmetic relationships of peak, peak-to-peak, average, and instantaneous values, and any other values as well, are applicable only to pure sine waves.

Effective Values of Sine Voltages and Currents

Measuring peak, average, and instantaneous voltages or currents of sine waves are simply attempts to assign a value to a changing waveform. It is true that the peak and peak-to-peak values are fixed, but these are only one or two points on the waveform, and while they are helpful, cannot be said to be truly indicative of the entire wave.

To get a more commonly used and more representative value, a comparison is made between AC and DC. A direct current flowing through a resistor produces heat. So does an alternating current. The effective value of an alternating sine current is that value which will produce the same amount of heat in a resistor as a direct current of the same numerical value. While the effective value of a sine wave can be obtained in this way experimentally, once it is obtained a numerical relationship can be used to move easily and arithmetically between the effective value and all the other sine wave values.

Figure 3-7 shows how to obtain the effective value. Each vertical line represents an instantaneous value of current with the lines selected at random, but the greater the number of lines the more accurate the answer. Each of these selected lines representing a voltage at some instant of time is measured, and then each voltage amount is squared, that is, multiplied by itself. These squared values are then added. The square root of the resulting addition supplies the effective voltage value, also known as the *root-mean-square value,* or *RMS*.

While the amount of a sine wave of voltage can be expressed in terms of peak, peak-to-peak, or instantaneous amounts, RMS is the value used most often.

The AC Sine Wave

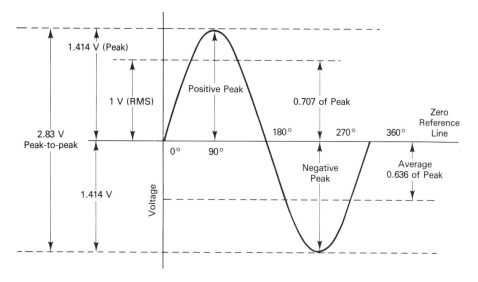

Figure 3-5. Dimensions of a sine wave.

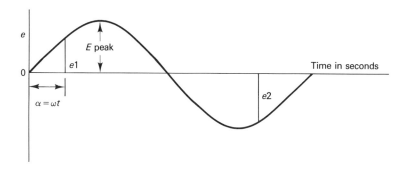

Figure 3-6. $e1$ and $e2$ are representative instantaneous values.

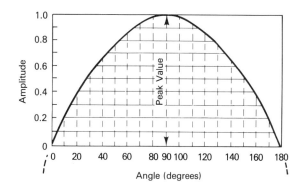

Figure 3-7. Effective value of a sine wave of voltage or current can be obtained from instantaneous values, shown here by vertical lines.

A data plate on an AC motor will show the required source voltage, and though this is an RMS, it is usually implied rather than stated.

For any current sine wave the effective value is

$$I_{\text{effective}} = 0.707 \times I_{\text{peak}}$$

The same formula is also applicable to sine waves of voltage:

$$E_{\text{effective}} = 0.707 \times E_{\text{peak}}$$

or

$$E_{\text{effective}} = \frac{E_{\text{peak}}}{1.414}$$

By assigning a value of 1 to the effective value,

$$E_{\text{peak}} = \frac{1}{0.707} = 1.414 \, E_{\text{effective}}$$

and in the case of peak-to-peak values,

$$E_{\text{peak-to-peak}} = 2 \times 1.414 \, E_{\text{effective}} = 2.828 \, E_{\text{effective}}$$

Note that 0.707 is the same as $0.5 \times \sqrt{2}$ and $1.414 = \sqrt{2}$.

The data plate mounted on the frame of a motor does not supply information about the AC other than to indicate the amount to be supplied to the motor input, and sometimes the frequency and the phase. The waveform of the voltage is a sine wave, unless otherwise specified, and the frequency is 60 Hz (i.e., 60 cycles per second, also shown as 60 cps), again unless otherwise specified.

The Variable RMS

The peak value of an AC voltage supplied by a power utility is a variable quantity, an amount that can decrease from its maximum value depending on how much of a load is being imposed. But since the RMS value is directly dependent on the peak value, this will also fluctuate. It can drop to 100 V, or even less, depending on load conditions. That is why motors and other electrically operated devices often indicate different amounts of line voltage. Some motors call for RMS line voltages such as 115, 117, 120, and 121 V.

In actual practice the RMS voltage will range from slightly under 100 to as much as 125. For a line voltage having an RMS value of 121 V, the peak voltage will be 171 and the peak-to-peak voltage will be twice this amount. However, the peak voltage will occur only twice during each cycle of the wave. The line voltage will also be zero twice during each cycle.

The fact that a motor is rated at 121 V RMS does not mean other voltage values can be ignored. In this example, 121 V is the amount required by various motor windings. These windings are insulated so as to separate their copper conductors from the iron core on which they are wound. The core is at ground or zero

The AC Sine Wave

potential. The insulation covering the wires must tolerate the peak voltage, and (as indicated above) for an RMS of 121 V will be 171 V peak.

Conversions

There are four ways of measuring a sine wave of voltage or current. Table 3-2 summarizes the relationships between these values.

Summary of Sine-Wave Voltage and Current Formulas

Average value of voltage or current sine wave

$$E_{average} = 0.637 \times E_{peak}$$

Peak value of voltage or current sine wave

$$E_{peak} = \frac{E_{average}}{0.637}$$

Peak value of voltage or current sine wave

$$E_{peak} = 1.57 \times E_{average}$$

Peak-to-peak value of voltage or current sine wave

$$E_{peak-to-peak} = 2 \times E_{peak}$$

Peak value of voltage or current sine wave

$$E_{peak} = E_{peak-to-peak}/2$$

Average value of voltage or current sine wave

$$E_{average} = E_{peak-to-peak} \times 0.3185$$

Average value of voltage or current sine wave

$$E_{average} = (E_{peak-to-peak}/2) \times 0.637$$

TABLE 3-2. SINE WAVE RELATIONSHIPS

Given this value of voltage or current	Multiply by this value to get			
	Average	Effective (RMS)	Peak	Peak-to-Peak
Average	—	1.11	1.57	1.274
Effective	0.9	—	1.414	2.828
Peak	0.637	0.707	—	2.0
Peak-to-Peak	0.3185	0.3535	0.50	—

Average value of voltage or current sine wave

$$E_{average} = E_{peak\text{-}to\text{-}peak} \times 0.5 \times 0.637$$

Average value of voltage or current sine wave

$$E_{average} = \frac{E_{peak\text{-}to\text{-}peak} \times 1}{2} \times 0.637 = E_{peak\text{-}to\text{-}peak} \times 0.5 \times 0.637$$

Peak-to-peak value of voltage or current sine wave

$$E_{peak\text{-}to\text{-}peak} = \frac{E_{average}}{0.3185}$$

RMS (effective) value of voltage or current sine wave

$$I_{effective} = 0.707 \times I_{peak}$$

RMS value of voltage or current sine wave

$$I_{effective} = \frac{I_{peak}}{1.414}$$

Peak value of voltage or current sine wave

$$I_{peak} = \frac{I_{effective}}{0.707}$$

Peak-to-peak value of voltage or current sine wave

$$E_{peak\text{-}to\text{-}peak} = 2.828 \times E_{effective}$$

RMS value of voltage or current sine wave

$$E_{effective} = \frac{E_{peak\text{-}to\text{-}peak}}{2.828} = \frac{E_{peak}}{1.414} = 0.707\, E_{peak}$$

PHASE

It is possible for a conductor, such as a copper wire, to carry a number of alternating currents of different frequencies or a combination of direct current and one or more alternating currents. It does not follow that all of the alternating currents must start and stop their individual cycles at the same time. If, as shown in Figure 3-8, there are two generators, A and B, supplying AC voltages, one generator may start before the other and so the sine waves they produce could be out of step.

This can be illustrated graphically, as in Figure 3-9. Here the two out-of-step voltages are produced by a pair of generators identified as E1 and E2. The separation in degrees of two voltages, or two currents, or possibly a voltage and a current, is called phase and is measured in degrees. The Greek letter theta, θ, is generally used to represent the phase angle. When voltages or currents start and

Phase

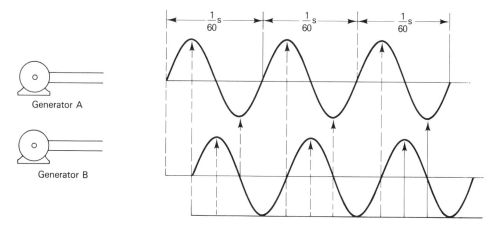

Figure 3-8. Phase separation of independent voltages controlled by separate generators.

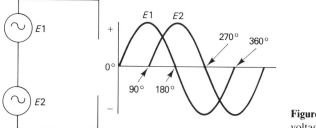

Figure 3-9. Voltage E1 leads voltage E2 by 90°.

stop at exactly the same time they are said to be *in phase;* if not, they are *out of phase.*

In the diagram of Figure 3-9 one of the voltages, E1, has started at 0°, while the other voltage, E2, starts at 90°. E2 does not start until E1 reaches its peak at 90°. Since E1 starts before E2, the action can be described by saying that voltage E1 leads E2 or that E2 lags E1.

Sometimes an alternator will contain an armature with just a single output voltage. It can be referred to as a *single-phase generator,* whose output is similar to the single-phase output waveform illustrated in Figure 3-10. This drawing shows three separate generators, A, B, and C. Each of the generators supplies a single-phase AC wave, but since the waves do not start and stop at the same time, they are out of phase. The same out-of-phase voltages can be obtained more economically by combining the three generators into one, and it can then be identified as such. The generator could have three individual armature coils mounted with an angular separation of 120°.

In some generators there are two armature coils positioned at right angles to each other. Known as a two-phase generator, it supplies two output voltages out

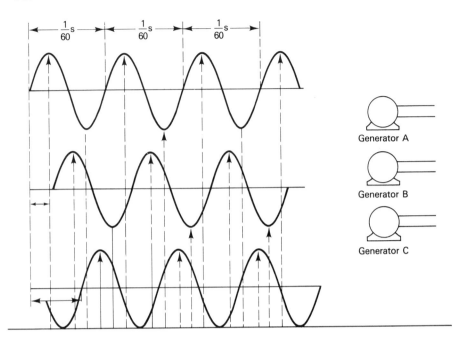

Figure 3-10. Out-of-phase voltages supplied by independent generators.

of phase with each other by 90°. However, three-phase AC generators are much more common. Known as polyphase generators, they can supply two or more voltages that can be used together or independently. Figure 3-11 shows two generators, although in actual practice both would be combined into one unit. Each phase of this generator supplies an AC potential of 120 V RMS, hence a pair of 120-V lines would be available. While Figure 3-11 shows a two-phase generator, the same thinking applies to a three-phase generator, except that with such a generator three 120-V lines are available.

Generator Frequency

When an AC generator is electromechanical, the output frequency is usually rather low. Most commonly the frequency is 60 Hz, although frequencies such as 25, 50, and 400 Hz are also used.

Generator Types

A generator can be electromechanical or electronic. The output frequency of the electromechanical type is within a fairly small range. An electronic generator can not only supply the same frequencies as the electromechanical type, but can have output frequencies measured in millions of hertz.

Reactance

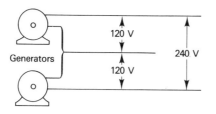

Figure 3-11. Independent generators with each having an output of 115 V. If the generators are in phase, the total output can be 230 V.

REACTANCE

When either a direct or alternating current flows through a resistor, a voltage is developed across that resistor based on the formula

$$E = I \times R$$

That voltage and current are in phase (Figure 3-12). Thus, $\theta = 0°$.

The Greek letter theta, θ, represents the phase angle, or the separation in degrees between the voltage and current. The same letter is used for the phase angle of several voltages or currents.

Inductive Reactance

If the resistor is replaced by a coil (an inductor) and an AC voltage is applied across the ends of the coil (Figure 3-13), that voltage and the resulting current flowing through the coil are out of phase. The separation between the two is a maximum of 90° but is usually somewhat less because of resistance inherent in the copper wire with which the coil is wound. The maximum phase angle is

$$\theta = +90°.$$

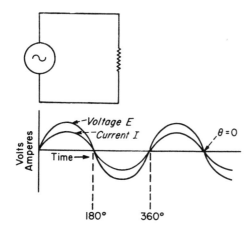

Figure 3-12. Voltage and current are in phase in resistive circuit.

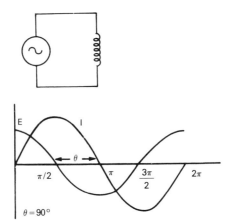

Figure 3-13. Current lags the voltage in an inductive circuit.

Assuming an ideal coil, that is, one that has no resistance, the opposition of the coil to current flow, known as inductive reactance, is

$$X_L = 2\pi f \times L$$

or

$$X_L = \omega L$$

where

$2\pi = 6.28,$

$f =$ the frequency in hertz,

$\omega = 2\pi f$ and

$L =$ the inductance of the coil in henrys.

Like resistance, reactance is measured in ohms.

In a more practical sense, the resistance of the coil is sometimes considered, and in that case the overall opposition of the coil to the flow of an alternating current is referred to as *impedance,* represented by the letter Z. Impedance is also measured in ohms. The impedance of the coil, including its resistance, is expressed as

$$Z = \sqrt{R^2 + X_L^2}$$

See Figure 3-14.

If the value of R is small compared to the reactance of the coil, it can be ignored. The smaller the value of R, the greater the phase angle between the voltage across the coil and the current flowing through it. Since the armature of a motor consists of a coil wound on an iron core, its inductive reactance is significant.

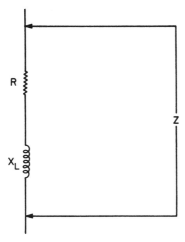

Figure 3-14. Series R-L circuit.

Capacitive Reactance

A capacitor is an important component in some types of AC motors. As in the case of inductors, a capacitor also has an opposition, measured in ohms, and known as *capacitive reactance*. The formula for calculating capacitive reactance is

$$X_c = \frac{1}{2\pi \times f \times C}$$

or

$$X_c = \frac{1}{\omega C}$$

where

X_c = the capacitive reactance in ohms,

C = the capacitance in farads,

f = the frequency in hertz,

$\omega = 2\pi f$, and

$2\pi = 6.28$.

Impedance

When a coil and capacitor are connected in series, both components will be involved in the control of current flow. The total opposition, measured in ohms, can be expressed as

$$Z = \sqrt{R^2 + (X_L - X_c)^2}$$

This formula assumes the value of X_L is greater than X_c. If X_c is the larger amount, transpose X_L and X_c. The formula then becomes

$$Z = \sqrt{R^2 + (X_c - X_L)^2}$$

OHM'S LAW FOR AC

Ohm's Law is as applicable for AC as it is for DC. Other than the substitution of impedance (Z) for resistance (R), the formulas are identical.

$$E = I \times Z$$

$$I = \frac{E}{Z}$$

$$Z = \frac{E}{I}$$

where Z is in ohms, E is in volts, and I is in amperes. However, Z is more complex than the simple R used in Ohm's Law for DC.

Table 3-3 is a summation of Ohm's Law formulas for AC. The unknown values of I, Z, E, and P are shown across the top, while the known values are listed in the column at the left. Table 3-4 is a summary of power and Ohm's Law formulas for AC and is a variation of Table 3-3.

POWER FACTOR

The fact that a motor is connected to an AC power line doesn't necessarily mean it will accept all the power delivered to it. The unused power will be returned to the voltage source, and for motors it is usually via the AC power line. Since the

TABLE 3-3. OHM'S LAW FORMULAS FOR AC

Known Values	Formulas for determining unknown values of...			
	I	Z	E	P
I & Z			IZ	$I^2 Z \cos\theta$
I & E		$\dfrac{E}{I}$		$IE \cos\theta$
I & P		$\dfrac{P}{I^2 \cos\theta}$	$\dfrac{P}{I \cos\theta}$	
Z & E	$\dfrac{E}{Z}$			$\dfrac{E^2 \cos\theta}{Z}$
Z & P	$\sqrt{\dfrac{P}{Z \cos\theta}}$		$\sqrt{\dfrac{PZ}{\cos\theta}}$	
E & P	$\dfrac{P}{E \cos\theta}$	$\dfrac{E^2 \cos\theta}{P}$		

TABLE 3-4. SUMMARY OF POWER AND OHM'S LAW FORMULAS FOR AC

Watts	Amperes	Volts	Impedance
$P =$	$I =$	$E =$	$Z =$
$I^2 R$	E/Z	IZ	E/I
$EI \cos \theta$	$\dfrac{P}{E \cos \theta}$	$\dfrac{P}{I \cos \theta}$	$\dfrac{E^2 \cos \theta}{P}$
$\dfrac{E^2 \cos \theta}{Z}$	$\sqrt{\dfrac{P}{Z \cos \theta}}$	$\sqrt{\dfrac{PZ}{\cos \theta}}$	$\dfrac{P}{I^2 \cos \theta}$
$I^2 Z \cos \theta$	$\sqrt{\dfrac{P}{R}}$	$\dfrac{\sqrt{PR}}{\cos \theta}$	$\dfrac{R}{\cos \theta}$

electrical power has made a complete round trip through the power lines, there will be some loss in the form of heat. This behavior of a motor with respect to power lines is referred to as power factor (*pf*) and can be expressed as

$$pf = \frac{E \times I \times \cos \theta}{E \times I}$$

or

$$pf = \cos \theta = \cos \frac{R}{Z}$$

where

$R =$ the resistance in ohms,

$Z =$ the impedance in ohms,

$E =$ the voltage in volts,

$I =$ the current in amperes,

pf = the power factor expressed as a decimal, and

cos = the cosine of the phase angle.

The maximum value of the power factor is 1 or unity, and is shown in the result as a percentage, by multiplying it by 100. Unity power factor is $1 \times 100 = 100\%$. A power factor of 100% is the most desirable working condition and means that all the power delivered by a power line has been accepted and utilized. For a purely resistive circuit the power factor is 100%, sometimes known as unity power factor.

POWER VERSUS ENERGY

There is a difference between power and energy, although the two terms are often used synonymously, and incorrectly so. Power is measured in watts, kilowatts, and megawatts. The time element is not a factor and does not appear in formulas in which watts are part of the calculations. The symbol for watts is the letter *P*, an abbreviation for power in watts.

Energy is measured in watt-hours and kilowatt-hours, and for motors, in horsepower-hours. The time element is always included. Electrical energy is always power multiplied by time. The basic unit is the watt-hour, indicating that a power of 1 W has been used for 1 hr. The symbol for energy is the letter *W*, an abbreviation for watt-hours. In terms of a formula,

$$W = P \times t$$

where

W = the energy expended in watts-hours,

P = the power in watts, and

t = the time in hours.

Power and Current

For a resistive circuit, power or current can be calculated by

$$P = I^2 \times R$$

$$I = \sqrt{\frac{P}{R}}$$

or in a reactive circuit—one containing capacitance or inductance, or both—by

$$P = I^2 Z \cos \theta$$

$$P = \frac{E^2 \cos \theta}{Z}$$

The formula $P = I^2R$ means a power factor of 100%, indicating that the load is completely resistive. This is not true of motors, since the armature and field coil windings are reactive elements. If, as an example, a motor has a power factor of 83%, then the power delivered to the motor must be multiplied by 0.83 to determine the actual amount of power used by the motor; 17% of the power, under this set of conditions, is returned to the power source.

For home use the power factor is close to 100%, since most home appliances are resistive—broilers, electric ranges, lamps, and so on. Some appliances, such as electric fans, turntables, cassette decks, and compact disk players, all of which use motors, do have a reactive component, but this is generally small in terms of power factor.

Since power is the product of voltage and current, the phase angle between voltage and current is a factor that must enter such calculations. In a DC circuit, power is simply $E \times I$. In a purely resistive AC circuit, power is also $E \times I$, since the phase angle, θ, is zero.

In an AC circuit containing a reactive element, and this is often the case for electric motors used in industry,

$$P = E \times I \times \cos \theta$$

where

P = the power in watts,

E = the voltage (in volts),

I = the current (in amperes), and

$\cos \theta$ = the power factor = the ratio of the resistance to the impedance, or R/Z.

No power is expended in a purely reactive circuit. Under these conditions, $\theta = 90°$, $\cos \theta = 0$ and $P = 0$. In any AC circuit the range is somewhere between 100% power factor and zero power factor, and so $\cos \theta$ is generally some value less than 1 and greater than 0.

SINGLE PHASE

The phrase *single phase* is applicable to AC motors and motor control circuits, for power, current, and voltage. A basic formula involving these values can be

$$\text{Watts} = \text{Voltage} \times \text{Current} \times \text{Power Factor}$$

In terms of symbols,

$$P = E \times I \times pf$$

where

P = the power in watts,

E = the voltage in volts,

I = the current in amperes, and

pf = the power factor expressed as a decimal or a percentage.

It must be in decimal form for use in problem solving. Similarly, E and I must be in basic units of volts and amperes; if they are in multiples or submultiples, they must first be converted.

Summary of Formulas for Single-Phase Circuits

$$P = E \times I \times pf$$
$$P = I^2 \times R$$
$$I = \frac{P}{E \times pf}$$

$$E = \frac{P}{I \times pf}$$

$$pf = \frac{P}{E \times I}$$

Voltage and Current Phase Relationships

It is possible for two or more voltages to be out of phase with each other or for voltages to be out of phase with currents. Similarly, a pair of currents can be in phase, or one current can lead or lag the other, as indicated in Figure 3-15.

To Find the Power When Voltage and Current are Known.

$$P = E \times I \times pf$$

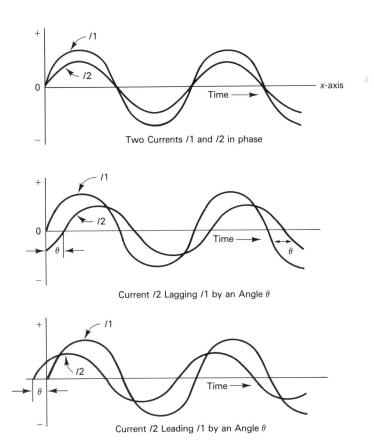

Figure 3-15. Currents can be in phase or one current can lead or lag the other.

Single Phase

Example:

An AC motor is connected to a 117-V, single-power line. Operating at a 75% power factor, the motor uses 5 A at full load. What is the motor's power requirement?

Solution:

$$P = E \times I \times pf$$
$$= 117 \times 5 \times 0.75$$
$$= 438.75 \text{ W}$$

To Find the Power When Current and Resistance Are Known.

$$P = I^2 \times R$$

Example:

A single-phase AC motor is connected to the power line by a 5-Ω resistor. How much power is used by this resistor when a current of 4 A flows through it?

Solution: Since no reactive components such as a coil or a capacitor are involved, the voltage across this resistor and the current through it are in phase. The power factor is 1 and thus can be omitted from the formula.

$$P = I^2 \times R$$
$$= 4 \times 4 \times 5$$
$$= 80 \text{ W}$$

Current

To Find the Current When the Power, Voltage, and Power Factor Are Known.

$$I = \frac{P}{E \cos \theta}$$

Example:

A motor is driving a load that requires a power input to the motor of 800 W. The single-phase AC line voltage is 121. The motor is rated as having a 90% power factor. What is the amount of current flow?

Solution:

$$I = \frac{P}{E \cos \theta}$$

$$= \frac{800}{121 \times 0.90}$$

$$= \frac{800}{108.9} = 7.35 \text{ A}$$

To Find the Current When the Power, Impedance, and Power Factor Are Known.

$$I = \sqrt{\frac{P}{Z \cos \theta}}$$

Example:

How much current does a 10-kW power load take from a single-phase power line if the circuit impedance is 65 Ω and its power factor is 65%?

Solution:

$$I = \sqrt{\frac{P}{Z \cos \theta}}$$

$$= \sqrt{\frac{10 \times 1{,}000}{65 \times 0.65}}$$

$$= \sqrt{\frac{10{,}000}{42.25}} = \sqrt{236.68} = 15.38 \text{ A}$$

To Find the Current When the Horsepower is Known in a Single-Phase Circuit

$$I = \frac{746 \times hp}{E \times \textit{eff} \times \textit{pf}}$$

Example:

A single-phase AC motor with a rated output of 2 hp is connected to a 120-V power line. It has an efficiency of 90% and a power factor of 70%. What is the amount of line current?

Solution:

$$I = \frac{746 \times hp}{E \times \textit{eff} \times \textit{pf}}$$

$$= \frac{746 \times 2}{120 \times 0.90 \times 0.70}$$

$$= \frac{1492}{75.6} = 19.74 \text{ A}$$

Voltage

To Find the Voltage When the Input Power, Current, and Power Factor Are Known in a Single-Phase Arrangement

$$E = \frac{P}{I \times \textit{pf}}$$

Single Phase

or

$$E = \frac{P}{I \cos \theta}$$

Example:

What is the input line voltage to a single-phase AC motor when its power requirement is 300 W? The current flow is 2 A and the motor's power factor is rated at 75%.

Solution:

$$E = \frac{P}{I \times pf}$$

$$= \frac{300}{2 \times 0.75}$$

$$= \frac{300}{1.5} = 200 \text{ V}$$

To Find the Voltage When the Power, Resistance, and Power Factor Are Known

$$E = \frac{\sqrt{PR}}{\cos \theta}$$

To Find the Voltage When the Power, Impedance, and Power Factor Are Known

$$E = \sqrt{\frac{PR}{\cos \theta}}$$

To Find the Power Factor in a Single-Phase Motor Circuit When the Input Power, Voltage, and Current Are Known

$$pf = \frac{P}{E \times I}$$

Example:

What is the power factor of a single-phase AC motor being supplied with an AC input power of 350 W and connected to a 121-V power line? The amount of current flow is 4 A.

Solution:

$$pf = \frac{P}{E \times I}$$

$$= \frac{350}{121 \times 4}$$

$$= \frac{350}{484} = 0.72 = 72\%$$

REAL POWER VERSUS REACTIVE POWER

Power that is used and not returned to the power source is called *real* or *true power*. Power returned to the source is known as *reactive power*. For real power only resistance (R) is involved:

$$P = I^2 \times R$$

For reactive power only inductive or capacitive reactance is involved:

$$P = I^2 \times X$$

where

P = the power in watts,

I = the current in amperes, and

X = either capacitive or inductive reactance (in ohms) or the vector sum of the two.

For apparent power involving reactance and resistance,

$$P = I^2 \times Z$$

The ratio of real power to apparent power supplies the power factor (*pf*):

$$pf = \frac{\text{real power}}{\text{apparent power}} = \frac{I^2 \times R}{I^2 \times Z} = \frac{R}{Z}$$

Wattmeters measure real power only. Power factor is a comparison between real power and apparent power.

Summary of Formulas Involving Power

$$P = \frac{E^2 \times \cos \theta}{Z}$$

$$P = I^2 \times Z \times \cos \theta$$

$$I = \sqrt{\frac{P}{E \times \cos \theta}}$$

$$I = \sqrt{\frac{P}{Z \times \cos \theta}}$$

$$I = \sqrt{\frac{P}{R}}$$

$$E = \frac{P}{I \times \cos \theta}$$

$$P = I^2 \times R$$

$$P = E \times I \times \cos \theta$$

$$P = \frac{E^2 \times \cos \theta}{Z}$$

$$P = I^2 \times Z \times \cos \theta$$

$$I = \frac{E}{Z}$$

$$I = \frac{P}{E \times \cos \theta}$$

$$I = \sqrt{\frac{P}{Z \times \cos \theta}}$$

$$I = \sqrt{\frac{P}{R}}$$

Average AC power in resistive circuit

$$P_{average} \; I^2 \times R$$

Peak AC power in resistive circuit

$$P_{peak} = I^2_{peak} \times R$$

Average AC power in resistive circuit

$$P_{average} = \frac{E^2_{average}}{R}$$

$$E = \sqrt{\frac{P \times Z}{\cos \theta}}$$

$$Z = \frac{E^2 \times \cos \theta}{P}$$

$$Z = \frac{P}{I^2 \times \cos \theta}$$

VA AND KVA

In the abbreviation *VA*, the letter *V* represents volts and *A* indicates amperes. VA is the multiplication of volts times amperes; that is, VA = V × A. In some motor installations the sum of the source voltages may be large, and so a multiple, KVA, is used. The letter *K* is used as a multiplier representing 1,000. Thus, 3 KVA is equivalent to 3,000 volts multiplied by the current in amperes. Unless otherwise indicated, VA and KVA both represent true power. An amount such as 2 KVA indicates that a true electrical power of 2 kilovolt-amperes is being utilized, not only by all the motors, but all other electrically operated components, including lights, heaters and so on.

VARS

The abbreviation *var* is used to distinguish reactive power from true power and represents the phrase *reactive volt-amperes.* A var is the reactive power of 1 reactive volt-ampere.

In some instances, where high amounts of power are involved, a kilovar is used instead. Thus, 1 kilovar = 1,000 vars. Vars can be expressed mathematically as

$$\text{var} = E \times I_{\text{reactive}}$$

A wattmeter is used for the measurement of AC power, but this is real power only, not reactive. A separate instrument, called an *electrodynamometer,* is used for measuring reactive power, but it can do so only for sine waves having a frequency for which the instrument is adjusted.

If a factory has an AC input power requirement of 175 A at 220 V, its power demand is $E \times I$ or 38,500 volt-amperes. If the factory has a total load that is practically resistive, the power factor is 100%, so what is delivered is true power. But if the power factor is some amount less than unity, possibly 85%, the KVA is still 38.5, but the real power is only $0.85 \times 38.5 = 32.725$ KVA or 85% of the delivered power. The difference between 38.5 KVA and 32.725 is reactive or unused power.

Table 3-5 supplies useful conversion information for determining current, power, KVA and horsepower when other data are supplied for DC motors and single-phase, two-phase, and three-phase AC motors.

Single-Phase KVA

To find kilovolt-amperes in single-phase circuits,

$$\text{KVA} = \frac{E \times I}{1{,}000}$$

Three-Phase KVA

To find kilovolt-amperes in three-phase circuits,

$$\text{KVA} = \frac{E \times I \times 1.732}{1{,}000}$$

TABLE 3-5. FORMULAS FOR DETERMINING AMPERES, HORSEPOWER, KILOWATTS, AND KVA

To find	Direct current	Alternating current		
		Single phase	2 Phase*—Four wire	Three phase
Amperes when Horsepower is known	$\dfrac{hp \times 746}{E \times \% \, eff}$	$\dfrac{hp \times 746}{E \times \% \, eff \times pf}$	$\dfrac{hp \times 746}{2 \times E \times \% \, eff \times pf}$	$\dfrac{hp \times 746}{1.73 \times E \times \% \, eff \times pf}$
Amperes when Kilowatts are known	$\dfrac{kW \times 1000}{E}$	$\dfrac{kW \times 1000}{E \times pf}$	$\dfrac{kW \times 1000}{2 \times E \times pf}$	$\dfrac{kW \times 1000}{1.73 \times E \times pf}$
Amperes when KVA is known		$\dfrac{KVA \times 1000}{E}$	$\dfrac{KVA \times 1000}{2 \times E}$	$\dfrac{KVA \times 1000}{1.73 \times E}$
Kilowatts	$\dfrac{I \times E}{1000}$	$\dfrac{I \times E \times pf}{1000}$	$\dfrac{I \times E \times 2 \times pf}{1000}$	$\dfrac{I \times E \times 1.73 \times pf}{1000}$
KVA		$\dfrac{I \times E}{1000}$	$\dfrac{I \times E \times 2}{1000}$	$\dfrac{I \times E \times 1.73}{1000}$
Horsepower—(output)	$\dfrac{I \times E \times \% \, eff}{746}$	$\dfrac{I \times E \times \% \, eff \times pf}{746}$	$\dfrac{I \times E \times 2 \times \% \, eff \times pf}{746}$	$\dfrac{I \times E \times 1.73 \times \% \, eff \times pf}{746}$

I = amperes; E = volts; % eff = percent efficiency; pf = power factor; kW = kilowatts; KVA = kilovolt-amperes; hp = horsepower.
*For three-wire, two-phase circuits the current in the common conductor is 1.41 times that in either of the other two conductors.

INPUT POWER (KVA) VERSUS OUTPUT POWER (HP)

For single-phase motors, the horsepower output can be calculated from

$$hp = \frac{\eta \times I \times E \times pf}{746}$$

or

$$hp = \frac{E \times I \times eff \times pf}{746}$$

and for three-phase motors

$$hp = \frac{\sqrt{3} \times I \times E \times \eta \times pf}{746}$$

where

hp = the output in horsepower,
η = the efficiency of the motor,
I = the line current,
E = the line voltage, and
pf = the power factor.

Example:

What is the KVA of a single-phase motor rated at 2,000 W? The input is 220 V AC and the current is 4 A.

Solution:

$$KVA = \frac{E \times I}{1,000}$$

$$= \frac{220 \times 4}{2 \times 1,000}$$

$$= 0.44 \text{ KVA}$$

Example:

What is the KVA of a three-phase motor whose source voltage is 440 V with a current flow of 2 A?

Solution:

$$KVA = \frac{440 \times 2 \times 1.732}{1,000}$$

$$= \frac{1524.16}{1,000}$$

$$= 1.524 \text{ KVA}$$

Example:

What is the horsepower output of a 220-V, single-phase motor if its current flow is 8 A, is 75% efficient, and operates at a power factor of 70%?

Solution:

$$hp = \frac{E \times I \times eff \times pf}{746}$$

$$= \frac{220 \times 8 \times 0.75 \times 0.70}{746}$$

$$= 1.24 \text{ hp}$$

Example:

Assuming that the motor in the above example was a three-phase instead of a single-phase type, what would be the effect on the horsepower output?

Solution: With no change in the input voltage, efficiency, and power factor, the only result would be an increase in the horsepower output of $\sqrt{3}$ or 1.732.

$$1.24 \times 1.732 = 2.15 \text{ hp}$$

THREE PHASE

If the AC input to a motor is from a polyphase power line, then this input is usually three-phase, since two-phase AC power is rare. Despite this there are two-phase motors with single-phase power converted to two-phase by a reactive component, such as a coil or capacitor, that is part of the motor and is external to the power line.

Power in Three-Phase Circuits

The power formulas for three-phase circuits are the same as those used for single phase except that the $\sqrt{3}$ factor, or 1.732, is involved.

In a single-phase resistive circuit,

$$P = E \times I$$

If the circuit is three-phase resistive, then

$$P = E \times I \times 1.732$$

Finally, if the three-phase circuit is reactive, that is, is either capacitive or inductive, then

$$P = E \times I \times pf \times 1.732$$

Although this looks like a new formula, it is the same used previously for power, but now taking *pf* and polyphase voltage into consideration.

Example:

A three-phase star-connected motor is wired to a 208-V circuit. It draws a current of 10 A and has a 50% power factor. How much input power is required by this motor?

Solution: Since the motor windings are inductive, the motor, as a load on the power supply, is regarded as a reactive element.

$$P = E \times I \times pf \times 1.732$$
$$= 208 \times 10 \times 0.50 \times 1.732$$
$$= 1{,}801.3 \text{ W} = 1.8 \text{ kW}$$

The power factor of a three-phase motor can be calculated from

$$pf = \frac{P}{\sqrt{3}(E_L)(I_L)}$$

or

$$pf = \frac{P}{E \times I \times 1.732}$$

where

pf = the power factor,

P = the power delivered to the motor,

E_L = the line voltage between any two phases, and

I_L = the line current.

Example:

What is the power factor in a three-phase system when the associated wattmeter indicates 3500 W, the AC voltmeter shunted across the AC power lines reads 220 V, and the ammeter pointer is at the 10-A mark?

Solution:

$$pf = \frac{P}{E \times I \times 1.732}$$
$$= \frac{3500}{220 \times 10 \times 1.732}$$
$$= \frac{3500}{3810.4} = 0.9185 = 91.85\%$$

Current in Three-Phase Circuits

The formula for finding the current in a three-phase circuit can be developed in the same way.

$$I = \frac{P}{E \times pf \times 1.732}$$

EFFICIENCY

Current in Three-Phase Circuits

Efficiency can be a factor in both AC and DC motors. If the horsepower is known in a single-phase circuit, the current can be calculated by

$$I = \frac{746 \times hp}{E \times eff \times pf}$$

This formula is the same as the formula for current for DC motors. The only difference is that in AC circuits the power factor must be included if it is less than unity.

Example:

An AC motor rated at 3 hp is connected to a single-phase 110-V AC power line. The efficiency of this motor is 90%. What are the current requirements of this motor, rated as having a 90% power factor?

Solution:

$$I = \frac{746 \times hp}{E \times eff \times pf}$$

$$= \frac{746 \times 3}{110 \times 0.9 \times 0.9}$$

$$= \frac{2{,}238}{89.1} = 25.1 \text{ A}$$

Efficiency is also a factor in electrical calculations for three-phase work. If the hp, operating voltage, efficiency, and *pf* are known, the amount of current can be calculated by

$$I = \frac{746 \times hp}{E \times eff \times pf \times 1.732}$$

This formula is similar to those used previously, except that the $\sqrt{3}$ factor is now involved in the calculations.

Example:

How much current is required by a 5-hp, 220-V, three-phase motor whose operating efficiency is 90% and power factor is 80%?

Solution:

$$I = \frac{746 \times hp}{E \times eff \times pf \times 1.732}$$

$$= \frac{746 \times 5}{220 \times 0.9 \times 0.8 \times 1.732}$$

$$= \frac{746}{44 \times 0.9 \times 0.8 \times 1.732}$$

$$= \frac{746}{54.9} = 13.59 \text{ A}$$

Voltage in Three-Phase Circuits

The voltage in a three-phase circuit can be calculated by transposing the voltage (E) and the current (I) in the formula shown above. The formula for the determination of the voltage becomes

$$E = \frac{746 \times \text{hp}}{I \times \text{eff} \times \text{pf} \times 1.732}$$

Example:

What is the amount of source voltage per phase for a 1-hp motor with an operating current of 4 A? The power factor of the motor is rated at 90%, and the efficiency is 85%.

Solution:

$$E = \frac{746 \times \text{hp}}{I \times \text{eff} \times \text{pf} \times 1.732}$$

$$= \frac{746 \times 1}{4 \times 0.9 \times 0.85 \times 1.732}$$

$$= \frac{746}{5.3} = 141 \text{ V}$$

To find the voltage in a high-efficiency three-phase circuit when the power and the current are known, use

$$E = \frac{P}{I \times \text{pf} \times 1.732}$$

Example:

A three-phase motor has a load that requires a motor power input of 1.3 kW. The total motor current is 8 A, and the power factor is 80%. Disregarding the efficiency, which is very high, what is the amount of input voltage?

Solution:

$$E = \frac{P}{I \times \text{pf} \times 1.732}$$

$$= \frac{1,300}{8 \times 0.8 \times 1.732} = 117 \text{ V}$$

Power Delivered to a Three-Phase Motor

The formula shown below is applicable to wye- and delta-wound stator windings.

$$P = \sqrt{3}(E_L)(I_L)\cos\theta$$

where

P = the total power delivered to the three-phase motor,

E_L = the line voltage measured between any two phases,

I_L = the current in a single line, and

θ = the angle between voltage and current in any phase.

Two-Phase Operation

The factor 2 is used as the constant in two-phase circuits, just as 1.732 is the constant for three-phase.

Two-Phase Motor Operating From a Three-Wire Line

It is possible for a two-phase motor to work from any two phases of a three-phase power line. The voltage relationships are expressed by

$$E_d = \sqrt{2}(E_p)$$

where

E_d = the voltage across the stator winding,

E_p = the voltage measured from the center tap of the stator winding to either side of the line, and

$\sqrt{2}$ = 1.414.

To find the power in a two-phase circuit, use

$$P = E \times I \times pf \times 2$$

Example:

How much power is being supplied to a two-phase AC motor that has a line current of 7 A and a 70% power factor? The line voltage is 220.

Solution:

$$P = E \times I \times pf \times 2$$
$$= 220 \times 7 \times 0.70 \times 2$$
$$= 2156 \text{ W} = 2.156 \text{ kW}$$

VOLTAGE TRANSFORMERS: STEP UP AND STEP DOWN

The ratio of secondary turns (N_s) to the primary turns (N_p) or the ratio of primary turns to the secondary turns of a transformer is known as the turns ratio, T_r. In terms of formulas, for a step-up transformer,

$$T_r = N_s/N_p$$

For a step-down transformer,

$$T_r = N_p/N_s$$

In either instance the turns ratio is a whole number. The amount of voltage step-up or step-down depends on the turns ratio. For a voltage step-up transformer,

$$T_r = E_s/E_p$$

For a voltage step-down transformer,

$$T_r = E_p/E_s$$

E_p is the voltage across the primary, E_s the secondary voltage. N_p is the number of primary turns; N_s the number of secondary turns. For the formulas shown above,

$$E_p \times N_s = N_p \times E_s$$

$$E_p = \frac{N_p \times E_s}{N_s}$$

$$E_s = \frac{E_p \times N_s}{N_p}$$

$$T_r = \frac{E_p \times N_s}{E_s}$$

$$N_s = \frac{N_p \times E_s}{E_p}$$

If information concerning the turns ratio is known, the actual number of primary and secondary turns isn't required. Thus,

$$E_s = T_r \times E_p$$

This formula can be rearranged to read

$$E_p = E_s/T_r$$

CURRENT TRANSFORMERS: STEP UP AND STEP DOWN

In transformers, the effects on current and voltage are inverse. A transformer will step down current by the same ratio that it steps up voltage. Conversely, a transformer will step up current by the same ratio that it steps down voltage.

Current Transformers: Step Up and Step Down

$$\frac{I_p}{I_s} = \frac{N_s}{N_p}$$

I_p and I_s represent primary and secondary currents. The current formula can be rearranged in the same manner as the voltage formula and in as many different ways.

$$I_p \times N_p = I_s \times N_s$$

then

$$I_p = \frac{I_s \times N_s}{N_p}$$

$$I_s = \frac{I_p \times N_p}{N_s}$$

$$N_p = \frac{I_s \times N_s}{I_p}$$

$$N_s = \frac{I_p \times N_p}{I_s}$$

chapter four

Formulas for Magnetic Circuits

Motors are involved in two types of characteristics: electric and magnetic. Those under the heading of electric include voltage, current, and resistance; those listed under magnetic have magnetic force, lines of flux, and reluctance. These characteristics are related, and it is necessary to be able to convert from one to the other. In some respects it is impossible to separate them. An electric current, direct or alternating, through a straight conductor or a coil, is accompanied by a magnetic field. Just as an electric current can encounter resistance, so too can lines of magnetic force encounter reluctance.

Electrical characteristics seem simpler, since they are more familiar. Voltage, current, and resistance are common terms, often part of daily living. Magnetic characteristics are not met as often, and furthermore, work under different systems.

CGS, MKS, AND ENGLISH SYSTEMS

Three systems involving magnetics are the CGS (centimeter-gram-second), the MKS (meter-kilogram-second), and the English or practical system. The CGS and MKS are metric; their common element is the second. While more than one system complicates the subject, there are practical reasons for all of them. We use MKS when the magnetic quantities are large, CGS when they are small, and English because of its familiarity.

Because of their differences, conversions must be made when necessary to utilize more than one system. But whichever system is selected, data must be consistently in that system. Table 4-1 lists CGS units; Table 4-2 lists MKS units.

CGS, MKS, and English Systems 137

TABLE 4-1. CGS UNITS

centimeter
oersted
gram
dyne
gilbert
erg
ergs per second
grams per cubic centimeter
dynes per square centimeter
centimeters per second
dynes per centimeter
gauss
second
maxwell

TABLE 4-2. MKS UNITS

meter
kilogram
joule
kilograms per cubic centimeter
newtons per square meter
webers per square meter
tesla (webers per square meter)
second
ampere-turns per meter
ampere-turns per weber
webers per ampere-turn

The MKSA System and SI Units

The MKSA is similar to the MKS. The MKSA (also called the Georgie) is an abbreviation for meter-kilogram-second-ampere units.

The International System of Units was established in 1960. It did not replace either CGS or MKS. The International System is an overall system with MKS considered as being included.

SI is the abbreviation for the International System of Units in all languages and is a modernized version of the metric system. In this system the hertz (Hz) is the SI unit of frequency equivalent to one cycle per second (cps). The meter/second is the SI unit of speed or velocity. The watt is the SI unit of power.

Standard Second

The recurrence of an alternating waveform is measured in time units, most often the *second*. A standard second is an extremely precise measurement of a unit of time. The standard second, used as a reference in time measurements, is the length of time required for 9,192,631,770 oscillations at the transition frequency of the cesium atom with a magnetic field of zero.

The English, or Practical, System

This system, known as the practical or English system, is also known as the U.S. system or the U.S. Customary System. The English system can be converted to other systems by changing its measurements to metric. No single system is suitable for solving all problems in magnetics.

ENGLISH AND METRIC UNITS

The English system uses English units of measurement, such as the inch, the square inch, the foot, the square foot, and magnetic units such as the ampere-turn. In a similar way, the CGS and MKS systems use metric units of measurement such as the centimeter and meter, and magnetic units such as the newton and joule. Table 4-3 is an aid for working back and forth between the English, CGS, and MKS systems. Table 4-4 is a listing of units in the three systems.

Commonly used abbreviations in the English system are in. for inches, ft for feet, and yd for yards. For the metric system, abbreviations are μm for micrometers, mm for millimeters, cm for centimeters, dm for decimeters, m for meters, dam for decameters, hm for hectometers, km for kilometers, and mym for myriameters. These are listed in Table 4-5.

The MKS and CGS units aren't directly interchangeable in a problem. If a problem involves MKS, all units must be in that system, and if not, must be converted. The same is true with CGS.

TABLE 4-3. CONVERSION FACTORS FOR MKS, CGS, AND ENGLISH UNITS

Multiply	by	To obtain
F in ampere-turns	$0.4\pi = 1.257$	F in gilberts
F in gilberts	$1/0.4\pi = 0.796$	F in ampere-turns
H in ampere-turns/in.	$0.4\pi/2.54 = 0.495$	H in oersteds
H in oersteds	$2.54/0.4\pi = 2.02$	H in amperes-turns/in.
B in maxwells/sq. in.	$1/6.45 = 0.155$	B in gauss
B in gauss B in maxwells/sq. cm.	6.45	B in maxwells/sq. in.
B in webers/sq. meter	10^4	B in gauss
B in gauss	10^{-4}	B in webers/sq. meter
B in maxwells/sq. in.	$10^{-4}/6.45 = 0.155 \times 10^{-4}$	B in webers/sq. meter
B in webers/sq. meter	6.45×10^4	B in maxwells/sq. in.
ϕ in maxwells ϕ in lines of flux	10^{-8}	ϕ in webers
ϕ in webers	10^8	ϕ in maxwells

F = magnetomotive force; H = field intensity; B = flux density; ϕ = flux

TABLE 4-4. CGS, MKS, AND ENGLISH UNITS

Term	Description	Symbol	CGS Units	MKS Units	English units	Notes
Flux	Total no. of lines	$\phi = \dfrac{\text{mmf}}{\mathcal{R}}$	1 maxwell = 1 line	1 weber = 10^8 lines	1 kiloline = 10^3 = 1,000 lines	Comparable to electric current
Flux density	Lines per unit area	$B = \dfrac{\phi}{A}$	1 gauss = 1 maxwell per cm^2	tesla = $\dfrac{\text{weber}}{m^2}$	kilolines per $in.^2$	$1\ m^2 = 10^4\ cm^2$; $1\ in.^2 = 6.45\ cm^2$
Magnetomotive force	Total force producing flux	$F = \phi \times \mathcal{R}$	gilberts = 1.256 × amp-turns	amp-turn	amp-turn	Corresponds to voltage, independent of length; 0.796 = 1/1.256
Field intensity, or magnetizing force	Force per unit length of flux path	H	1 oersted = 1 gilbert per cm	amp-turn per m	amp-turn per in.	Corresponds to voltage per unit length; 1 in. = 2.54 cm
Reluctance	Opposition to flux	$\mathcal{R} = \dfrac{\text{mmf}}{\phi}$	gilbert or maxwell	amp-turn per weber	amp-turns per kiloline	Corresponds to resistance
Permeability	Ability to concentrate flux	$\mu = \dfrac{B}{H}$	gauss or oersted	$1.256 \times 10^{-6} \mu^*$	$3.2 \times 10^{-3} \mu^*$	μ of air or vacuum is 1

* Multiply μ by these factors to calculate B from H in MKS or English units.

TABLE 4-5. LINEAR MEASUREMENTS IN ENGLISH AND METRIC

1 micrometer	=	0.001 mm
1 micrometer	=	0.000001 m
1 millimeter	=	0.0393700 in.
1 millimeter	=	0.00328 ft
1 centimeter	=	10 mm
1 centimeter	=	0.393700 in.
1 centimeter	=	0.032808 ft
1 centimeter	=	0.01093611 yd
1 meter	=	39.3700 in.
1 meter	=	3.280833333 ft
1 meter	=	1.09361 yd
1 decimeter	=	10 cm
1 decimeter	=	3.937 in.
1 meter	=	10 dm
1 meter	=	100 cm
1 meter	=	1,000 mm
1 decameter	=	10 m
1 decameter	=	393.7 in.
1 hectometer	=	10 dam
1 hectometer	=	328 ft, 1 in.
1 kilometer	=	10 hm
1 kilometer	=	0.62137 mi
1 myriameter	=	10 km
1 myriameter	=	6.2137 mi
1 inch	=	25.40005 m
1 inch	=	2.540005 cm
1 inch	=	0.02540005 m
1 foot	=	304.8006 mm
1 foot	=	30.48006 cm

Since both MKS and CGS are metric, any measurements involving the English system must first be converted to metric if either MKS or CGS are to be used. Table 4-6 is a more detailed arrangement of metric to English and English to metric units.

FLUX LINES

All magnets, permanent or electro, are accompanied by magnetic lines of force, also known as *flux*. These lines are continuous and unbroken, forming a complete loop around a straight wire carrying a current, and can be represented by continuous lines surrounding the wire. When the current stops the magnetic lines collapse and disappear. If the current is increased, the number of magnetic lines increases, forming larger loops.

The current flow can be AC or DC. With DC the magnetic lines remain steady; with AC the magnetic lines are alternately weak and strong. The electro-

OHM's Law for Magnetic Circuits

TABLE 4-6. METRIC TO ENGLISH AND ENGLISH TO METRIC

	Metric Units To English Equivalents		
Lengths	1 millimeter	0.03937	in.
	1 centimeter	0.3937	in.
	1 meter	39.37	in. or 1.0936 yd
	1 kilometer	1093.61	yd or 0.6214 mi
Areas	1 square millimeter	0.00155	in.^{2}
	1 square centimeter	0.155	in.^{2}
	1 square meter	10.764	ft^2 or
		1.196	yd^2
	1 square kilometer	0.3861	mi^2
Volumes	1 cubic millimeter	0.000061	in.^{3}
	1 cubic centimeter	0.061	in.^{3}
	1 liter	61.025	in.^{3}
	1 cubic meter	35.314	ft^3 or
		1.3079	yd^3
	English System Units To Metric Equivalents		
Lengths	1 inch	25.4	mm or
		2.54	cm
	1 foot	0.3048	m
	1 yard	0.9144	m
	1 mile	1.6093	km
Areas	1 square inch	645.16	mm^2
		6.452	cm^2
	1 square foot	0.0929	m^2
	1 square yard	0.8361	m^2
	1 square mile	2.59	km^2
Volumes	1 cubic inch	16,387.2	mm^3
		16.3872	cm^3
	1 cubic foot	0.02832	m^3
	1 cubic yard	0.7646	m^3

magnetic field has its polarity determined by the direction of current flow. With DC that polarity remains fixed, since its current always moves in the same direction. With AC the magnetic strength not only increases and decreases, but since this current changes its direction, the polarity of the magnetic field reverses accordingly.

OHM'S LAW FOR MAGNETIC CIRCUITS

In magnetic circuits flux is comparable to current in electric circuits, magnetomotive force (F) to voltage, and reluctance to resistance. It can be expressed as

$$\text{Magnetic lines of force} = \frac{\text{magnetomotive force}}{\text{reluctance}}$$

The formula for the calculation of flux can be manipulated more easily if it is written with the help of symbols

$$\phi = \frac{F}{\mathcal{R}}$$

where

ϕ = the flux,

F = the magnetomotive force in ampere-turns, and

$\mathcal{R}$ = the reluctance in rels.

Ampere turns can be changed to gilberts by multiplying by 0.4π or 1.257. The formula for flux can also be written as

$$\phi = \frac{IT}{\mathcal{R}}$$

where I is the current in amperes and T is the number of turns.

Known as the Ohm's Law for magnetic circuits, it can be rearranged if the magnetomotive force or the reluctance are the unknown values. It will then appear as

$$F = \phi\mathcal{R} \quad \text{and} \quad \mathcal{R} = \frac{F}{\phi}$$

The unit of reluctance in the MKS system is the *ampere-turn per weber*. This can be written as

$$\mathcal{R} = \frac{F}{\phi} \text{ ampere-turns per weber}$$

This equation can be transposed to read

$$\phi = \frac{F}{\mathcal{R}} \text{ webers}$$

and

$$F = \phi\mathcal{R} \text{ ampere-turns}$$

Example:

A solenoid wound with 4,000 turns of wire has a reluctance of 3 rels. When the solenoid is activated it carries a current of 4 A. How many lines of magnetic flux surround the coil? How many lines of flux exist inside the coil form of the solenoid?

Solution: Using Ohm's Law for magnetic circuits,

$$\phi = \frac{IT}{\mathcal{R}}$$

Substituting the values supplied,

$$\phi = \frac{4 \times 4000}{3} = 5{,}333 \text{ (approximately)}$$

The number of lines of flux is the same either inside the coil form or outside it.

THE MAXWELL

In the CGS system each single magnetic line called flux or a magnetic line of force is a maxwell. A more common unit is the weber (pronounced *vayber*), since it is used to describe a large number of maxwells. The weber is part of the MKS system. Maxwells can be converted into webers, or vice versa by

$$1 \text{ weber} = 10^8 \text{ maxwells} = 100{,}000{,}000 \text{ maxwells}$$

$$1 \text{ maxwell} = 10^{-8} \text{ weber} = 1/100{,}000{,}000 \text{ weber}$$

The maxwell (the single line of flux in the CGS system) and the weber (in the MKS having a value of 100,000,000 maxwells) are extremes. A more practical unit is the microweber (μweber) having a value of 100 lines or 100 maxwells. In the English system a kiloline, or 1,000 lines, has a greater utility than a single line.

The unit of flux density in the MKS system is the weber per square meter. One weber is equal to 10^8 maxwells, and 1 m^2 equals 10^4 cm^2, thus 1 weber/m^2 is equal to 10^4 gauss.

Table 4-7 lists the characteristics of magnetic lines (lines of flux).

TABLE 4-7. CHARACTERISTICS OF MAGNETIC LINES

1. All magnetic lines are continuous, unbroken. There are no partial magnetic lines.
2. All magnetic lines have the same direction: away from the north pole toward the south pole.
3. Magnetic lines around a straight current-carrying conductor are circles. The circles are just one of various geometric shapes that can be assumed by the lines, depending on the position and shape of nearby ferrous materials.
4. Magnetic lines that "leave" a magnet continue through the space surrounding the magnet via the north pole and "re-enter" the magnet via the south pole, continuing through the body of the magnet from the south pole to the north pole.
5. Magnetic lines do not "flow;" they occupy a greater space external to the magnet than in the body of the magnet itself.
6. Magnetic lines are crowded together where the magnetic field is strong and are farther apart where the field is weak.
7. Magnetic lines do not cross each other.
8. Magnetic lines repel each other sideways.
9. Magnetic lines are capable of passing through non-ferrous substances. They are not hindered by non-metallic materials such as paper, glass, wood, and cloth.
10. Because ferrous materials such as steel and iron have a reluctance much lower than air, magnetic lines pass through such materials much more easily than through air.
11. Unlike magnetic poles attract each other; similar magnetic poles repel.
12. Magnetic lines are often represented by either straight or curved lines equipped with an arrow. The arrow does not indicate motion, but simply the assumed direction of the lines.

TABLE 4-7. CHARACTERISTICS OF MAGNETIC LINES *(continued)*

13. The magnetic lines produced by a permanent magnet and those resulting from an electromagnet are identical.
14. Magnetic lines have no weight. A conductor carrying a current will have surrounding magnetic lines. That conductor weighs the same, with or without current flow.
15. All currents, whether alternating or direct, are accompanied by magnetic lines.
16. A permanent magnet suspended at its center and free to rotate will have its north pole pointing toward the earth's north magnetic pole. The magnet's north pole is sometimes called a *north-seeking pole,* but is actually the magnet's *south pole.*
17. The magnetic lines surrounding a permanent magnet will have their number decreased if the magnet is heated or if it is struck vigorously, such as with a hammer.
18. If a permanent magnet is cut in half, the result is a pair of magnets with each having its own north and south poles. Further division of the magnet will result in a series of smaller magnets, each equipped with a north and a south pole.
19. The north pole of a permanent magnet will be attracted by the south pole of an electromagnet. The behavior of the magnet lines of a permanent magnet is similar to the behavior of the lines of an electromagnet.
20. The total number of lines external to either a permanent or an electromagnet is referred to as a *field.* The stronger the magnet the greater the number of lines in that field. A field can be represented by straight or curved lines, with each of the lines headed by an arrow. In some instances a magnetic line is represented by a short, broken line. This does not mean the magnetic line is broken but is represented this way so as to simplify the drawing.
21. The strength of the magnetic field around a coil depends on the total number of turns, the amount of current, and the inductance of the coil. An increase in any of these factors will strengthen the magnetic field. The inductance of the coil can be increased by using a ferrous material such as iron or steel as the core. In some coils the core is fixed in position; in others it is adjustable. By moving the core in and out of the coil the strength of the magnetic field around the coil can be varied.
22. The strength of the magnetic field around a straight conductor carrying a current is small. That strength can be increased by winding that straight conductor into the form of a coil.
23. Magnetic lines tend to take the shortest path.
24. Adjacent magnetic lines in opposite directions attract.
25. Magnetic lines tend to shorten, behaving somewhat like rubber bands. Because of this behavior they are sometimes referred to as *lines of force.*
26. While similar poles repel and like poles attract, they must be close enough to interact.

MAGNETIC FORCE AROUND A CONDUCTOR

When a direct current flows through a straight conductor, it is accompanied by a steady magnetic field which surrounds that conductor. The strength of the magnetic field at any point along that conductor can be found by

$$H = \frac{2I}{10d}$$

where

$I =$ the current in amperes,

$H =$ the strength of the magnetic field in oersteds, and

$d =$ the distance from the conductor in centimeters.

MAGNETIC FORCE BETWEEN POLES

When dissimilar poles of a pair of bar magnets are brought close to each other, they will attract. If the poles are similar, they will repel. The force, whether one of attraction or repulsion, will be inversely proportional to the square of the distance between them, measured in centimeters, and directly proportional to the product of the strengths of the two poles. In terms of a formula,

$$F = \frac{M_1 \times M_2}{D^2}$$

where

F = the force,

M_1 and M_2 = the strengths of the two poles, and

D = the distance between the poles.

M_1 and M_2 are in terms of unit poles. A unit pole is one that can exert a force of 1 dyne on an identical pole in air or in a vacuum, when the distance between the poles is one centimeter. The dyne is the unit of force in the CGS system.

FORCE ON A CONDUCTOR

When a conductor carries a current, it becomes an electromagnet. When that conductor is positioned in a magnetic field the force acting on that conductor is expressed by

$$F = \frac{8.85\ BIl}{10^8}$$

where

F = the force in pounds,

B = the flux density in lines per square inch,

I = the current in amperes, and

l = the length of the conductor in inches.

The data supplied is in the English system.

Example:

A conductor that is 10 in. long carries a current of 8 A. The conductor is positioned at right angles to 80,000 flux lines per square inch. What is the amount of force exerted on the conductor in pounds?

Solution:

$$F = \frac{8.85 \times 80{,}000 \times 8 \times 10}{10^8}$$

$$= \frac{8.85 \times 8 \times 10^4 \times 8 \times 10^1}{10^8}$$

$$= \frac{566.4}{10^3} = 0.5664 \text{ lb}$$

Essentially the same formula can be used in the MKS system, appearing as

$$F = BlI$$

where

B = the flux density in webers per square meter,

l = the length of the conductor in meters,

I = the current in amperes, and

f = the force in newtons.

THE DC MAGNETIC FIELD

The current (produced by a DC voltage) passing through a wire rises to a peak, remains constant for a time controlled by an on-off switch, and then drops to zero when that switch is opened. For the time of its duration it is actually a pulse voltage. An electric current is always accompanied by lines of magnetic flux, which develop practically instantaneously to a peak amount, remain constant for the time duration of the current, and then collapse during the same time of current stoppage. The polarity of the magnetic field does not change, since the magnetic lines of flux remain in the same direction.

THE AC MAGNETIC FIELD

Assuming a sine wave of current, its accompanying magnetic field will increase sinusoidally from zero to a peak when the current reaches its 90-deg. point and will then gradually decrease to zero at 180 deg. Here both the current and the magnetic field will be zero. The current will then increase to a maximum again, but because of the polarity reversal of the applied voltage, it will flow in the opposite direction. Because of this current change, the magnetic field will reverse its polarity.

The behavior of a magnetic field produced by an alternating current depends on the waveform of that current and the kind of load through which that current will flow. Thus, a voltage having a trapezoidal waveshape placed across an inductor will result in a surrounding magnetic field having a sawtooth shape.

Shape of Magnetic Lines

Like the circumference of a wheel, magnetic lines of flux cannot be said to have a beginning or an end. Though they are circular when surrounding a straight wire carrying a current, they can assume not only a circular shape, but a distorted circular, a loop, an ellipse—the shapes, however, are not necessarily perfect geometric shapes. Magnetic lines can be made to expand or contract, but do not have a "flowing" motion typical of an electric current. Adjacent magnetic lines repel each other. There is a limit to the number of lines of flux that can be accommodated by an iron bar of given dimensions, and when that point is reached the bar is said to be *saturated*. Theoretically, the number of lines in the space surrounding a coil carrying a current has no limit, since its existence depends on the inductance of the coil and the amount of current passing through it. The magnetic lines surrounding or contained in either a permanent or electromagnet, that is, the sum of all the lines, is known as a *magnetic field*.

FLUX DENSITY

Flux density is the number of magnetic lines of force per unit area. In the CGS system that area is the square centimeter, with the lines of flux perpendicular to it. A single line of flux per square centimeter is known as a *gauss* and is represented by the capital letter B. When the word gauss is used to describe a line of flux it is understood that the area is 1 cm². Thus a reference to 100 B indicates that 100 lines of magnetic force are passing through an area of 1 cm². A magnet used for holding papers to a refrigerator door has a strength of 100 B.

In terms of a formula,

$$\phi = B \times A$$

The total amount of flux, ϕ, is equal to the product of the area, A, in square centimeters multiplied by the number of gauss. Flux density, B, is equal to the lines of flux, divided by the area through which the lines pass and is represented by

$$B = \frac{\phi}{A}$$

Flux density is not constant for any particular magnet. The number of lines in the body of a magnet and extending out of and around that magnet does not change, assuming no external force. But the cross-sectional area is smaller inside the body of the magnet than outside. This means the flux density is greater inside the magnet than external to it. The standard MKS unit of flux density is the *tesla* or webers per square meter. There are 10,000 gauss per tesla.

The measurement of flux density, usually in the air surrounding a magnet, makes use of another term, the *oersted*. A field intensity of 100 oersteds is the same as 100 lines per square centimeter. Like the gauss, the oersted is a unit in the CGS system. While it may seem strange to have two different terms for flux density, the oersted is more commonly associated with electromagnets, thus using the oersted as a unit generally implies a temporary magnet, that is, an electromagnet.

Example:

A bar magnet has a flux of 5,000 maxwells measured at right angles to these lines. The end of the magnet where the measurement was made is a square measuring 8 cm on one side. What is the flux density in gauss?

Solution:

$$B = \frac{\phi}{A} = \frac{5000}{8 \times 8} = \frac{5000}{64} = 78 \text{ gauss/cm}^2$$

A line of flux produced by a permanent magnet and an electromagnet are different only in that the magnetic line of a permanent magnet is constant, while that of an electromagnet can be made variable. The lines of flux of a permanent magnet have a long lifetime, assuming the magnet is not subjected to some external magnet force, is heated, or struck. The lines of flux of an electromagnet are completely dependent on the current flowing through the conductors. These magnetic lines will collapse when the current flow is stopped.

RELUCTANCE

The opposition to the existence of magnetic lines in a material is known as *reluctance* and is comparable to resistance in an electric circuit. The symbol for reluctance is the script letter $\mathcal{R}$; the unit of reluctance is the *rel*. Prior to the adoption of the term *rel*, the unit was known as the CGS unit of reluctance. For an electromagnet, reluctance is usually measured in ampere-turns per weber.

The reluctance of a magnetic circuit, or any part of that circuit, is proportional to its length, inversely proportional to a selected cross-sectional area, and inversely proportional to the permeability of the material. The *permeability* is the ease with which magnetic lines can exist in a substance. The relationship of reluctance and permeability is expressed mathematically as

$$\mathcal{R} = \frac{l}{A\mu}$$

where

$\mathcal{R}$ = the reluctance,

l = the length in centimeters,

A = the cross section in square centimeters, and

μ = the permeability.

Reluctance

Since this formula involves metric units, a conversion must be made if a reluctance problem is supplied in the English system.

Example:

A magnet in the form of a cylinder has a diameter of 3 in., a length of 8 in. and is made of steel having a permeability of 1,300. What is the reluctance of this magnet?

Solution: A first step is to convert the diameter and length from units in the English system to metric: 3.0 in. = 7.62 cm., 8.0 in. = 20.3 cm. The cross-sectional area is

$$A = \frac{\pi(7.62)^2}{4} = 45.6 \text{ cm}^2$$

$$\text{The area of a circle} = \frac{\pi (D)^2}{4}$$

Calculating the reluctance supplies:

$$\mathcal{R} = \frac{20.3}{45.6 \times 1,300} = 0.000342 \text{ rel}$$

The unit of reluctance is based on a cubic centimeter of air, that is, a cubic centimeter of air is assigned a standard reluctance of 1 rel.

If we assume there were 3 cm^3 of air between the north and south poles of a permanent magnet, with the volume of air equivalent to three cubes each having a length, width, and height of 1 cm, each of these cubes would have a reluctance of 1 rel. Since there are three cubes, the total reluctance is 3 rels.

Reluctances in Series

Reluctances in series can be expressed by

$$\mathcal{R}_t = \mathcal{R}_1 + \mathcal{R}_2 + \mathcal{R}_3 + \mathcal{R}_4 + \cdots \text{ rels}$$

If the reluctance is not known directly it can be calculated from the length, cross-sectional area, and the permeability of each section. The series reluctance would then be

$$\mathcal{R}_t = \frac{l_1}{A_{1}\mu_1} + \frac{l_2}{A_{2}\mu_2} + \frac{l_3}{A_{3}\mu_3} + \frac{l_4}{A_{4}\mu_4} + \cdots \text{ rels}$$

The reluctance of air to magnetic lines is substantially greater than that of a material such as iron. If two rectangular magnets are arranged so that the north pole of one magnet faces the south pole of the other with a slight separation between the two, the total reluctance will be determined by the air gap, since it will be so much greater than that of the combined reluctances of the two magnets.

Figure 4-1 shows a horseshoe magnet whose ends are bent so the north pole of the magnet faces its south pole. There are four sections of this arrangement with each having its own reluctance, with these identified as $\mathcal{R}_1$, $\mathcal{R}_2$, $\mathcal{R}_3$, and $\mathcal{R}_4$. The lines of magnetic flux exist in each of four sections. The reluctance of the curved and straight portions of the magnet is considerably less than that of the section

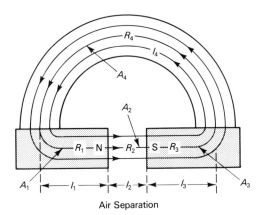

Air Separation

Figure 4-1. Reluctances in series.

consisting only of air between the magnets. The total reluctance can be calculated by using the formula for series reluctances, but this total will be very close to the value of $\mathcal{R}_2$, the air between the poles.

Reluctances in Parallel

Reluctances can be connected not only in series but in parallel as well. The formula used is similar to those for resistors wired in shunt.

$$\frac{1}{\mathcal{R}_t} = \frac{1}{\mathcal{R}_1} + \frac{1}{\mathcal{R}_2} + \frac{1}{\mathcal{R}_3} + \frac{1}{\mathcal{R}_4} + \cdots$$

Another formula for reluctances in parallel is

$$\mathcal{R}_t = \frac{\mathcal{R}_1 \mathcal{R}_2}{\mathcal{R}_1 + \mathcal{R}_2}$$

When reluctances are joined in parallel, the total resulting reluctance is less than that of any of the individual reluctances. If three identical reluctances are in parallel, the overall reluctance is one-third of any of the units.

PERMEABILITY

In an electric circuit, the reciprocal of resistance is conductance and is expressed as $G = 1/R$, where G is the conductance in siemens (abbreviated S; formerly called the *mho*) and R is the resistance in ohms. In a magnetic circuit, permeability (also known as permeance) is the reciprocal of reluctance and is supplied as

$$\mathcal{P} = \frac{1}{\mathcal{R}}$$

Permeability

Permeability is represented by the script letter $\mathcal{P}$, though in some instances an italic letter is used. Permeability is equal to magnetic flux divided by magnetomotive force. It is also defined as the ratio of the flux density in gausses or in lines per square centimeter B, to a magnetizing force, H, in oersteds or in gilberts per square centimeter. In terms of a formula, it is

$$B = \mu H$$

where

B = the flux density,

μ = the permeability, and

H = the magnetizing force.

The tesla is sometimes used as the unit of flux density.

The permeability of a magnetic material, however, is not constant. This is because the number of flux lines produced by a magnetizing force, such as ampere-turns, varies, as indicated in Figure 4-2. From 0 to point A on the graph the number of flux lines rises rapidly with a linear increase in the magnetizing force. From point A to point B, however, the number of lines no longer increases as rapidly; and from point B fewer lines are produced, even though the magnetizing force keeps rising constantly. The maximum permeability shown in the graph is 5,000, but at this point it shows symptoms of leveling off. Because of this behavior, quite often (but not always) permeability is given as a range, from approximate minimum to approximate maximum. Table 4-8 supplies permeability figures for some commonly used magnetic materials.

Permalloy, Perminvar, and Sendust are the trademark names for special types of magnetic alloys. Permalloy is 78.5% nickel and 21.5% iron; Perminvar consists of 45% nickel, 30% iron, and 25% cobalt.

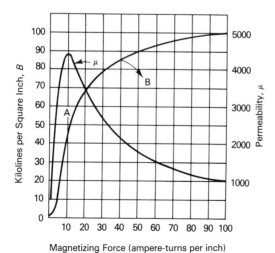

Figure 4-2. The number of flux lines does not increase linearly with an increase in the magnetizing force.

TABLE 4-8. APPROXIMATE PERMEABILITIES OF MAGNETIC MATERIALS

Cobalt	170
Iron-cobalt alloy	13,000
Iron, commercial annealed	6,000 to 8,000
Nickel	400 to 1,000
Permalloy	over 80,000
Perminvar	2,000
Sendust	30,000 to 120,000
Silicon steel	5,000 to 10,000
Steel, cast	1,500
Steel, open-hearth	3,000 to 7,000

Relative Permeability

This is a comparison of the permeability of a substance to air or a vacuum, both of which are assumed to have a value of 1. Permeability is identified by the Greek letter mu, μ. However, unlike other magnetic quantities, such as reluctance (rel), it does not have a unit. The permeability of a material has a number not followed by any identification.

RELUCTIVITY VS PERMEABILITY

The reluctance of a conductive substance is directly proportional to its length and inversely proportional to its cross-sectional area. In terms of a generalized expression this can be written as

$$\mathcal{R} = \frac{l}{A}$$

where l is the length in centimeters, and A is the cross-sectional area in sq. cm.

By itself this form is inadequate for it does not take into consideration that for equal volumes substances have different amounts of opposition to the establishment of lines of magnetic flux, known as reluctivity.

Since reluctivity (also called *specific reluctance*) is the reciprocal of permeability, it can be expressed as

$$\nu = \frac{1}{\mu} \quad \text{or} \quad \mu = \frac{1}{\nu}$$

where μ is the permeability and ν is the reluctivity. Reluctivity supplies an indication of the reluctance of a substance per unit volume, with its unit in rels per cubic centimeter. The range of reluctivity extends from approximately 0.000008333 to 0.00588.

Permeance

Just as reluctance shows itself as opposition to the establishment of magnetic lines, so too is permeability the opposite, the ability of a material to allow the existence of flux lines. Because reluctance and permeability have an inverse relationship it is possible to use either in the solution of magnetic problems. The permeability of a vacuum or air is considered as unity.

The reluctivity of magnetic materials has a wide range, as shown in Table 4-9. Since working with decimals can be difficult, it is preferable to use permeability figures instead. Table 4-9 lists the same substances as in Table 4-8. The listing in Table 4-9 is obtained by dividing the numbers in Table 4-8 into 1.

PERMEANCE

Permeability, the ease with which a magnetic field can be established in a ferrous material, is the reciprocal of reluctance. In a magnetic circuit, permeability is comparable to conductance in an electric circuit. The unit used for permeability in the MKS system is webers per ampere. It can also be considered as magnetic flux divided by magnetomotive force. In the CGS system it is equal to the magnetic flux in maxwells divided by the magnetomotive force in gilberts. In terms of a formula in the MKS system it can be stated as

$$\mathcal{P} = \frac{\phi}{F} \text{ webers per ampere-turn}$$

also

$$\mathcal{P} = \frac{1}{\mathcal{R}} \text{ weber per ampere-turn}$$

Conductance in electric circuits and permeability in magnetic circuits are comparable. Both conductances and permeabilities are added directly to find the total. The total permeability of conductors in parallel is:

$$\mathcal{P} = \mathcal{P}_1 + \mathcal{P}_2 + \mathcal{P}_3 + \mathcal{P}_4 + \cdots$$

TABLE 4-9. RELUCTIVITY OF MAGNETIC SUBSTANCES

Cobalt	0.00588
Iron-cobalt alloy	0.0000769
Iron, commercial annealed	0.0001666 to 0.000124
Nickel	0.0025 to 0.001
Permalloy	0.0000125
Perminvar	0.0005
Sendust	0.00003333 to 0.000008333
Silicon steel	0.0002 to 0.0001
Steel, cast	0.0006666
Steel, open-hearth	0.000333 to 0.000142857

MAGNETOMOTIVE FORCE

In an electrical circuit voltage is also referred to as electromotive force, or EMF. For voltage the corresponding quantity in a magnetic circuit is magnetomotive force, abbreviated as mmf and represented by F. There are other comparisons between electric and magnetic circuits, as indicated in Table 4-10.

For an electromagnet, the force is represented by the ampere turns, NI. This consists of the current in amperes, I, multiplied by the number of turns of wire, N. Magnetomotive force is that force that results in the production of lines of flux. The formula for calculating mmf involves a constant, 0.4π, where π is equal to 3.1416. The constant is 0.4π or $0.4 \times 3.1416 = 1.257$. The complete MKS formula is

$$F = 0.4\pi NI = 1.257 NI$$

where

F = the mmf in gilberts,

π = a constant equal to 3.1416,

N = the total number of wire turns of a coil, and

I = the current in amperes.

The gilbert, representing magnetomotive force, is a unit in the CGS system and is the force required to produce 1 maxwell of flux in a magnetic circuit having a reluctance of 1 rel. The gilbert is also equal to about 0.8 ampere-turn. The mmf in gilberts = $10/4\pi$ ampere-turns. The gilbert per centimeter is the practical CGS unit of magnetic intensity. Gilberts per centimeter are the same as oersteds.

To convert F in ampere-turn to F in gilberts, multiply ampere turns by 0.4π, or 1.257. If the data are supplied in gilberts they can be changed to F in ampere-

TABLE 4-10. COMPARISON OF ELECTRIC AND MAGNETIC CIRCUITS

	Electric circuit	Magnetic circuit
Force	Volt, E, or EMF	Gilberts, F, or mmf
Flow	Ampere, I	Flux, ϕ, in maxwells
Resistance	Ohms, R	Reluctance, $\mathcal{R}$, or rels
Law	Ohm's Law, $I = \dfrac{E}{R}$	Rowland's Law, $\phi = \dfrac{F}{\mathcal{R}}$
Intensity of force	Volts per cm of length	$H = \dfrac{1.257 IN}{l}$, gilberts per cm of length
Density	Current density—for example, amperes per cm^2	Flux density—for example, lines per cm^2, or gauss. The telsa is sometimes used as the unit of flux density.

turns by multiplying F in gilberts by $1/0.4\pi$ or 0.796. (These conversions are given in Table 4-3.)

THE AMPERE-TURN

The ampere-turn, NI, is magnetic field strength per unit length, and is represented by H. This representation of magnetic field strength can be stated in ampere-turns per inch (the English system) or in oersteds. The oersted is the unit of magnetic field strength in the CGS electromagnetic system. The oersted replaced the gauss for this purpose by international agreement in 1930.

To convert from the English system where H is supplied in ampere-turns per inch, to H in oersteds, multiply ampere-turns per inch by $0.4\pi/2.54 = 0.495$. To obtain H in ampere-turns/inch when the data are oersteds in multiply by $2.54/0.4\pi = 2.02$.

In the MKS system,

$$H = \frac{F}{l} \text{ ampere-turns per meter}$$

MANIPULATING MAGNETIC FORMULAS

There are three ways of manipulating magnetic formulas: by transposition, by cross-multiplication, and by substitution. In all instances the objective is to get the unknown value by itself. Transposition is applicable to formulas that have three values, cross-multiplication to those that have four, and substitution to those that have any number.

A representative three-value formula is

$$\mathcal{R} = \frac{F}{\phi} \text{ ampere-turns per weber}$$

This formula can be transposed to read

$$\phi = \frac{F}{\mathcal{R}} \text{ webers}$$

or

$$F = \phi \mathcal{R} \text{ ampere-turns}$$

The formula for flux supplied earlier,

$$\phi = \frac{F}{\mathcal{R}} \text{ webers}$$

can not only be rearranged by transposition but by substitution as well. However,

$$F = IN \text{ ampere-turns}$$

and

$$\mathcal{R} = \frac{l}{\mu_r \mu_v A} \text{ ampere-turns per weber}$$

by substituting both of these in the formula for flux, it becomes

$$\phi = \frac{\mu_r \mu_v INA}{l} \text{ webers}$$

This formula can be modified to supply an answer in terms of flux density, again by making a substitution. Thus, from

$$B = \frac{\phi}{A} \text{ webers per square meter}$$

we can obtain

$$B = \frac{\mu_r \mu_v IN}{l} \text{ webers per square meter}$$

The formula can be manipulated further:

$$F = IN = Hl \text{ ampere-turns}$$

Substituting for IN its value Hl and simplifying gives

$$B = \mu_r \mu_v H \text{ webers per square meter}$$

Example:

A current of 200 mA flows through a coil having 300 turns. What is the value of the magnetomotive force in ampere-turns? In gilberts?

Solution:

$$F = IN \text{ ampere-turns}$$
$$= (200 \times 10^{-3})(300)$$
$$= (200)(300)(10^{-3})$$
$$= 60{,}000 \times 10^{-3} = 60 \text{ ampere-turns}$$

To convert F in ampere-turns to gilberts multiply it by 0.4π, or 1.257.

$$60 \times 1.257 = 75 \text{ gilberts}$$

(Note the conversion of milliamperes to amperes: 200 mA = 200×10^{-3} ampere.)

MAGNETIC AND NON-MAGNETIC MATERIALS

Materials can be categorized in a number of ways, one of which is how they react to a magnetizing force. The categories can be listed as ferromagnetic, paramagnetic, and diamagnetic.

Ferromagnetic

The main ferromagnetic materials are iron, steel, nickel, cobalt, and their alloys. They are characterized by their ability to become magnetized and demagnetized more than other metallic substances. Their behavior in this regard varies from one to the other. They are used to form permanent magnets and as cores for electromagnets. In this latter application the result is a stronger magnetic field.

The direction of the magnetic force (Figure 4-3) induced in a ferromagnetic material is in the same direction as the applied magnetic field.

Paramagnetic

Using air or a vacuum as a reference standard of unity, paramagnetic substances have a permeability that is slightly greater and which is independent of the magnetizing force. Representative paramagnetic materials include platinum, chromium, and manganese.

Diamagnetic

These are materials that have a permeability that is slightly less than air or vacuum. Bismuth is representative, but other substances in this category include mercury, silver, gold, and copper. Copper, widely used in the manufacture of inductors (coils), does not exhibit any retention of magnetism, even though a current flowing through the coil can produce a strong magnetic field around it. If the coil contains an iron core, that core can retain enough magnetism to be identified as a permanent magnet. The coil itself does not have this characteristic. The difference is that the core is ferromagnetic; the copper wire is not.

When a diamagnetic material is subjected to the influence of a magnetic force, the resulting effect is that it becomes magnetized in a direction opposite that of the magnetizing force, as indicated in Figure 4-4.

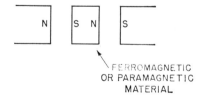

Figure 4-3. Ferromagnetic or paramagnetic materials become magnetized in the same direction as an external magnetizing force.

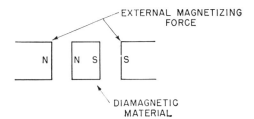

Figure 4-4. Diamagnetic material is magnetized in a direction opposite that of the magnetizing force.

MAGNETIC AGING

There are two basic aspects concerning the aging of a magnet. For an air-core inductor magnetic aging isn't applicable, since the field around such a coil is dependent only on the number of turns and the strength of the current. When the current is stopped, the coil no longer has magnetic characteristics.

For a permanent magnet, the component may be deliberately aged, or the aging process takes place naturally. For some permanent magnets aging generally refers to a loss of magnetic strength due to a gradual change in their metallurgical composition. Accelerated aging can be caused by an increase in ambient temperature or by subjecting the magnet to vibration or physical shock.

In some instances the magnetic strength of a magnet is deliberately lowered so that ultimately the number of lines of flux reach a constant level. The reduction in magnetic strength indicates a decrease in permeability.

Core loss can be a significant factor for both AC and DC motors, since they basically consist of iron-core coils. Both the field coils and the armature coils are wound on iron cores. An increase in core loss means a reduction in the electromagnetic fields surrounding these coils.

Core-Loss Aging Coefficient

This refers to a test procedure in which the core is heated at 100°C for 600 hours, following which the core loss is expressed as a percentage.

The Curie Point

The magnetization of a ferromagnetic material is temperature-dependent. With an increase in temperature, the substance will lose its ferromagnetic properties. The temperature at which it will do so is referred to as the Curie point. Conversely, as the temperature is lowered, a substance will indicate magnetization properties, and the point at which it will do so is also known as the Curie point.

MAGNETIC SATURATION

A ferrous material can be magnetized, but the process cannot be continued indefinitely. An increasing magnetomotive force applied to a ferrous substance will not result in a linear increase in the development of lines of flux but will reach a level at which there will be no further lines. The ferrous material is then said to be magnetically saturated. The saturation point can often not be maintained upon the removal of the applied magnetomotive force.

REMANENCE

When a ferrous material is subjected to a magnetizing force that is subsequently removed, some of the flux will decrease in quantity. The amount of flux remaining is referred to as remanent magnetism or remanence. Remanent magnetism may be subjected to an aging process so as to produce a permanent magnet whose field will remain constant over a long time.

RESIDUAL INDUCTION

The removal of a magnetizing force, as in the case of remanent magnetism, does not necessarily mean that the magnetizing force is zero at that time, but can be any value above zero. Residual magnetic induction, sometimes simply called *residual induction*, is the amount of flux remaining when the magnetizing force is reduced to zero.

If the magnetizing force is produced by a symmetrical AC waveform, such as a sine wave, the magnetizing force passes through zero twice during each cycle. This means that the iron core of a surrounding coil also reaches its residual induction at the same time.

RETENTIVITY

If a ferrous material is subject to a magnetizing force such that the material reaches magnetic saturation, the amount of flux remaining in the material following removal of the magnetizing force, no matter what the strength of that force might be, is referred to as *retentivity*. The retentivity of a substance that has been magnetized can be reduced to zero through the application of a negative magnetic field. A negative magnetic field is one whose polarity opposes that of the field because of the retentivity of the magnet. The term *coercivity* identifies the reverse magnetic force used to reduce magnetic retentivity to zero.

COERCIVE FORCE

Coercive force, identified as H_c, is the amount of magnetomotive force required to reduce the retentivity of a magnetic substance to zero.

HYSTERESIS

Some of the characteristics involved in the magnetization and demagnetization of a ferromagnetic material can be depicted graphically as in the graph in Figure 4-5, known as a hysteresis curve, hysteresis loop, or *B-H* curve. Basically, this graph demonstrates the tendency of a ferromagnetic material to retain its magnetism.

The graph starts at 0, the intersection point of a pair of straight lines, one representing the magnetizing force, *H*, the other a vertical line representing flux density, *B*. It is assumed that at point 0 the ferromagnetic material is completely neutral, that is, completely unmagnetized and has no magnetic lines of flux, as from some previous treatment. As a magnetizing force is applied, the flux density of the material increases until point a.

At point a, the magnetizing force is gradually reduced and the curve from a to b shows the effect of this reduction. When the magnetizing force reaches zero, the ferromagnetic material still has a substantial flux density, as indicated by point b on the flux density line. The material being magnetized has become a weaker magnet, but not by much. Because of the retentivity of the iron, the remanence or residual magnetism can be said to be substantial.

To reduce the remanence to zero, the polarity of the magnetizing force must be reversed, and when this reversed magnetizing force is applied the curve will move in the direction represented by b-c. At point c the magnetizing force is negative, and the ferromagnetic material is completely unmagnetized.

If this negative magnetizing force is increased, the ferrous material will become magnetized once again, but with reverse polarity; its poles, north and south, will become transposed. As the magnetizing force is increased in strength, the flux density will increase once again, until point d is reached. The only difference in the magnet between points a and d is in the reversal of the magnet's polarity.

If the reverse magnetizing force is now gradually reduced in strength, the flux density will decrease, as shown in the graph, from point d to e. At point e the

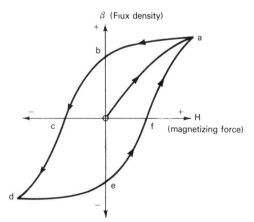

Figure 4-5. Hysteresis loop.

reverse magnetizing force is zero but the ferrous material is still magnetized. The difference between points b and e is only a matter of polarity. The north and south poles of the magnet have become interchanged, but in both cases the flux density has remained the same.

The polarity of the magnet can be changed once again by now applying a positive magnetizing force, as indicated by the curve from e to f. At point f the magnet has become completely demagnetized, but if the magnetizing force is increased the ferrous material will gradually become magnetized until point a is reached once again.

Note that hysteresis is due to the application of a magnetizing field produced by an AC source. This isn't applicable to that resulting from a non-varying direct current, for here there is no magnetic field reversal.

HYSTERESIS LOSS

No motor is 100% efficient, and the mechanical energy output is never equivalent to the electrical energy input. This is because of energy losses due to friction, I^2R losses in the motor armature and field coils, and hysteresis. All these losses show themselves in the form of heat.

The hysteresis curve indicates that magnetization lags behind the magnetizing force. It requires energy for a magnetic material to move through a complete cycle of magnetization and demagnetization. The amount of energy required is proportional to the area of the hysteresis loop. Hysteresis loss is also dependent on the kind of ferromagnetic material used, and is greater for some than for others. Hysteresis loss can be calculated from

$$\rho = kvf B_{max}^x$$

where

ρ = power loss in watts,
k = a constant for a given specimen,
v = volume of iron in cubic meters,
f = frequency in hertz,
B = flux density in tesla, and
x = an index between 1.5 and 2.3, often taken as 2.

chapter five

DC Motors

Compared to AC, there are relatively few DC motors, so-called since their source voltage is DC. The power for such motors can be supplied by batteries (wet, gel, or dry types) or the output of a power supply connected to an AC power line. One of the great advantages of such motors is their independence from fixed-position voltage sources such as an outlet, and so they are used extensively in all vehicles: cars, buses, recreational vehicles, boats, and planes. They are also used in portable electronic equipment, including compact disk players, tape decks, and turntables. In the home, even when an AC source is readily available, battery-operated DC motors are conveniently free from any need for a connection to an AC outlet. Aside from differences in source voltage, DC and AC motors have many similarities, and in at least one instance, the universal motor, either AC or DC can be used as the electrical power input.

RIGHT-HAND MOTOR RULE

Figure 5-1 illustrates the basic operation of a motor using a DC power source. The current from this source is controlled by a resistor and is allowed to pass through a single wire copper conductor, positioned in the path of a flux field supplied by a permanent magnet. Because of the reaction between the magnetic field supplied by the current flowing through the conductor and the field of the magnet, the conductor will tend to move upward.

The reaction between the two magnetic fields is represented by the outstretched fingers of the right hand. The thumb points upward to indicate the di-

The Commutator

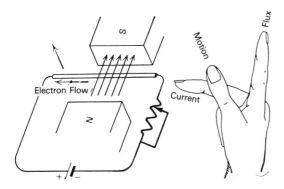

Figure 5-1 The right-hand motor rule.

rection of movement of the conductor. The forefinger shows the direction of the magnetic field of the permanent magnet, and the remaining finger points in the direction of flow of the battery current.

The listing in Table 5-1 is that of the various kinds of DC motors.

The fact that a motor is listed separately in Table 5-1 does not mean it is radically different from all the others. Thus a split-field series-wound is just a modified series-wound type.

There is still one other type of motor, but it is not included in this listing, since it is independent of any power source, either AC or DC. This is the air motor, which works from compressed air, and is thus in a category of its own.

THE COMMUTATOR

Fundamentally, a motor consists of the reaction between a pair of magnets, one of which is fixed in position, the other mounted on a shaft to permit rotation. The fixed-position magnet can be either a permanent type or an electromagnet. The rotating electromagnet, or armature, presents a problem, and that problem is how to make a connection to it while it is in motion. This is done by a commutator and

TABLE 5-1. TYPES OF DC MOTORS

The series-wound motor
The split-field series-wound motor
The shunt-wound motor
The compound motor
The differentially-wound motor
The cumulative-compound motor
Stepless motors
Permanent-magnet motors
Brushless DC motors
Stepping motors
Hybrid-stepping motors
Universal (DC and AC)

brushes. Located at the end of each commutator bar are slits or risers into which the leads from the armature coils are soldered.

The commutator consists of a number of copper segments in rectangular form. These copper segments are insulated from each other and are mounted in circular fashion near one end of the shaft of the motor, but insulated from that metal shaft. Each of the bars is the terminal of a coil, with two bars for each coil and with each of the coils referred to as an *armature*. Since each commutator is mounted on the shaft, the commutator and the shaft rotate together. The rotation of the shaft subjects the commutator to centrifugal force and so tends to move the copper segments outward. To overcome this they are kept in place by clamping rings that have a wedge-like shape. These rings are held together by bolts and nuts.

The copper segments composing the commutator are made of hard-drawn or drop-forged copper separated by thin layers of mica, an insulating material. The thickness of the mica ranges from 0.02 to 0.05 in., depending on the dimensions of the commutator segments (Figure 5-2).

Figure 5-3 shows one way of representing the commutator segments and the armature coils. The number of commutator elements will vary from one DC motor to the next, ranging from about 30 on small DC motors to as many as several hundred for larger machines.

It is difficult to represent the commutator in its circular form. A more practical method is to view the commutator as flattened, as in Figure 5-4. This method of presentation is known as a plan view and it is used not only for the commutator but for the armature as well.

Cleaning the Commutator

Clean the commutator with fine-grade sandpaper or commutator polishing paste. Do not use emery cloth, since its particles can short the commutator bars.

BRUSHES

Brushes are solid materials made of carbon and held in place by a brush holder, with a pair of brushes for each armature coil. While brushes are most commonly made of carbon, other substances such as copper gauze or copper graphite are used for low-voltage motors. Carbon brushes can be categorized as hard carbon, electrographitic, graphite, and metal graphite.

Brush Materials

A listing of materials used for brushes is supplied in Table 5-2.

Hard carbon. Brushes of the hard-carbon type are strongly abrasive, have high resistance, a relatively high voltage drop, and have a low current-carrying ability. On the plus side, they are able to tolerate vibration, physical shock, and

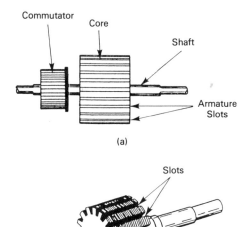
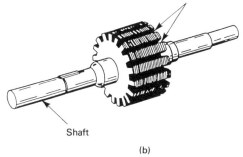
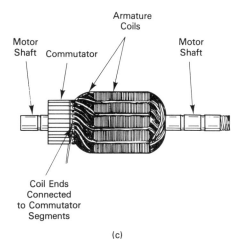

Figure 5-2 Armature core and commutator prior to insertion of armature windings into core slots (a); armature laminations assembly on motor shaft (b); completed armature assembly with armature coils connected to commutator segments (c).

voltage reversal. Because of their structure they are able to keep the commutator clean.

Electro-graphitic. The electro-graphitic type of brush is noted for having a high graphite content. In comparison with hard carbon brushes, they have a more moderate resistance and hardness, and a higher current-carrying capacity. They have moderately low friction and supply good lubrication, and thus are preferred when working with high commutator speeds.

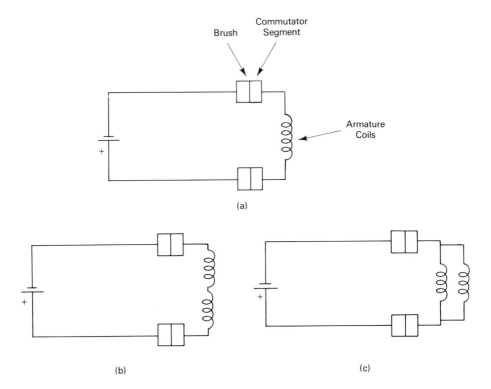

Figure 5-3 Current is delivered to armature coil when brushes rest on commutator segments (a); armature coils can be in series (b) or in parallel (c).

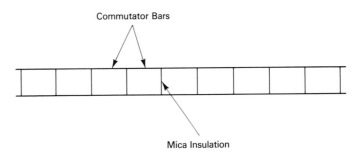

Figure 5-4 Plan view of commutator segments.

TABLE 5-2. BRUSH MATERIALS

Hard Carbon
Electro-graphitic
Graphite
Metal Graphite

Graphite. Graphite brushes have low hardness and low resistance. They also have a medium level of abrasiveness and voltage drop, and a high current-carrying capacity. They supply good lubrication and are desirable for high-speed motors.

Metal Graphite. This brush, a combination of metal and graphite, has low resistance and hardness. It also has a low voltage drop. The abrasiveness depends on its metal content. It has a higher current-carrying capacity than other types of brushes and is characteristically used with low-voltage motors.

Brush Selection

Motor manufacturers may make recommendations for the brush type to use with specific models. Generally, the selection of a brush depends on the level of source voltage, the amount of voltage drop (IR drop) that can be tolerated, the maximum current to be carried by the brushes, operating conditions, and whether the motor is to be used in continuous or intermittent operation. Brush wear is more serious when the motor must be started and stopped frequently, but is also dependent on the motor type. Thus, high-voltage shunt DC motors operated intermittently will exhibit greater brush wear. If a motor is subject to vibration and shock, a harder-than-usual carbon brush may be required. If operational quietness is needed, a softer carbon brush may be indicated.

Brushes are current conductors and work as a current link between the source voltage and the motor armature. Thus they must be capable of carrying the maximum current required by that winding. The amount of voltage applied to the armature is equal to the source voltage minus the voltage drop across the brushes.

Electrolytic Action

Brushes are polarized in the sense that one or more brushes are connected to the negative terminal of the voltage source, with the other brushes wired to the positive terminal. The negative brush shows greater wear, with the amount of wear differing from one type to the other. To equalize the wear factor, different brush types are used in some instances for the positive and negative connections.

Brush Holders

A brush holder (Figure 5-5) is a device for holding a brush in the correct position with respect to the armature and at the proper tension. For inexpensive motors used with home appliances, the holder can be nothing more than a small metallic rectangular container. The container and the brush are considered as integrated units, thus both must be replaced when the brush is worn.

Brush holders are available in two basic types: those that are adjustable and those that are non-adjustable. Adjustable means that the holder and brush can be

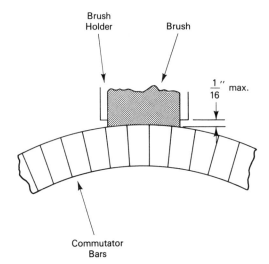

Figure 5-5 The brush may contact one or more commutator segments.

moved closer to or farther away from the commutator and that the holder and brush can be moved either to the left or right. The brushes are also adjustable to allow the maximum surface area of the brush to be kept in contact with the commutator segments. Depending on the motor, the brush can be positioned to obtain left- or right-hand rotation of the armature.

The end of the brush must make full contact with the commutator, with that part of the brush known as the heel facing in the direction of rotation. The opposite end is known as the toe. The direction of rotation of the commutator must be against the toe, as shown in Figure 5-6. Brushes do not simply rest against the commutator but are usually spring-loaded so as to supply a pressure against the commutator. The pressure ranges from 1.75 to 2.5 psi (pounds per square inch) for graphite types and up to 5 psi for copper/graphite brushes. To obtain the correct pressure, the spring tension should be adjustable. Even with brush pressure the brushes must be able to move freely in their holders. To be able to make good electrical contact with the source voltage and to avoid adding electrical resistance to the brushes, the upper part of the brush is copper plated and is connected to the brush holder through the use of a pigtail. A pigtail is used to permit the connection to "ride" with any brush movement during motor operation or brush adjustment. The pigtail is constructed of twisted copper wire. The brush holder itself must not make contact with the commutator but should ride from 1/16 to 1/8 in. above its surface.

THE ARMATURE

The armature consists of a number of coils mounted on a shaft, with the coil ends soldered to specific commutator bars. The armature is actually an iron-cored coil, the core being made of laminations having a thickness of approximately 0.014 in., although this dimension does not apply to all motors (Figure 5-7). Laminations are

The Armature

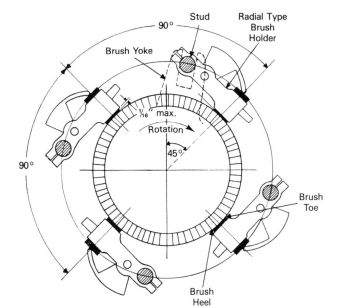

Figure 5-6 Assembly using four brushes. Direction of rotation must be against toe of brush.

used to minimize the development of eddy currents. The laminations are separated by a fine coating of varnish on their surfaces.

The armature is constructed with slots into which the armature coils are placed. The active part of the armature coil is its horizontal section that fits into a slot. Sometimes a single conductor is used for each slot, and in some motors two are used. The slots can be skewed (Figure 5-8[a]) or straight (Figure 5-8[b]).

The iron core of the armature is cylindrical, with the armature coils wound around the core and fitted into slots on the outer horizontal surface. Depicting the armature in three dimensional (Figure 5-9) helps give an idea of what the armature looks like, but does not allow one to trace the connections of the armature coils to the commutator. To do that it is necessary to view the arrangement as though it were on a flat surface. Unfortunately, such drawings often appear quite complicated (see Figure 5-10).

Number of Turns per Armature Coil

The number of turns for each armature coil may be shown as a single winding, but each coil may have 30 turns or more, and may be shown as a single, double, or

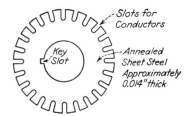

Figure 5-7 Annealed sheet steel lamination for armature.

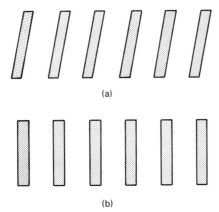

Figure 5-8 Armature slots can be skewed (a) or straight (b).

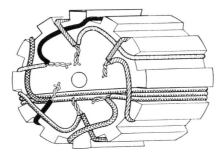

Figure 5-9 Three-dimensional view of armature coils positioned in armature slots.

triple turn, as in Figure 5-11. However, there is at least one start and one finish lead wired to each commutator bar.

Figure 5-12 is a symbol representation of an armature coil used when drawing motor circuits. To save space in a drawing and to avoid confusion, it is drawn as a diamond with a start lead and an end lead. These are the leads connected to commutator bars. The number of coil turns is usually not specified. The two vertical lines represent that part of the coil that is positioned in the armature slots. The slanted rear part of the drawing indicates that part of the coil that is vertical and occupies the rear portion of the armature core. The slanted front portion is that part across the front of the core.

The Armature's Magnetic Field

The current supplied by a DC source moves into the armature coils via brushes sliding over commutator segments. The current enters the armature coil and flows through it, and in so doing produces a magnetic field around it. The armature, then, is an electromagnet. The laminations of which the armature core is made are iron, because of its high permeability. The armature coils are wound so that when one side of the coil is under a south pole of a field coil, the other side is under a north pole. The side of a coil is its horizontal length (sometimes called an element)

The Armature

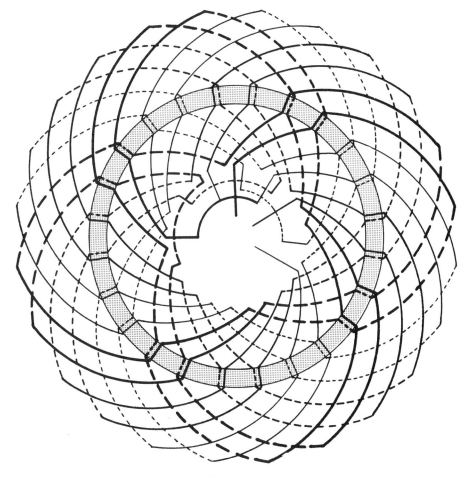

Figure 5-10 Armature winding.

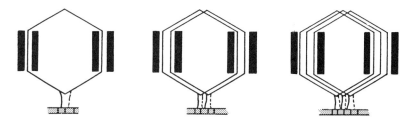

Figure 5-11 Single-, double-, and triple-turn armature coils. The solid line leading to the coil is the winding start; the dashed line is the finish.

Figure 5-12 Symbol for an armature coil. The number of coil turns is indeterminate unless otherwise indicated.

and is that part of the coil that fits into a slot. The shape of the coil is somewhat rectangular rather than round.

One horizontal coil element is fitted into one slot; the opposite horizontal side into another slot. One of the sides is at the bottom of a slot; the other side at the top. In this way each slot is occupied by two different coils. An armature coil may have a single turn or a number of them.

Partially surrounding the armature is another magnet known as the *field*. This magnet can be a permanent type or another electromagnet. It is the reaction between the armature's magnetism and the field's magnetism that produces rotation of the armature.

TYPES OF WINDINGS

Various types of armature windings are used to help secure desired motor operating characteristics. Table 5-3 supplies a listing. The terms are not exclusive and can be combined. Thus a lap winding could also be a simplex, progressive, symmetrical, single element type. Other combinations are also possible. Basically, though, there are just two fundamental types of armature windings: lap and wave. The others

TABLE 5-3. TYPES OF ARMATURE WINDINGS

Lap
Simplex lap
One-, two-, three-, and four-element simplex lap
Duplex lap
One-, two-, and three-element duplex lap
Four-element triplex lap
Wave
Simplex wave
Progressive
Retrogressive
Symmetrical

Types of Windings

are modifications. Coil windings differ in the ways in which their start and finish leads are connected to the commutator bars.

Lap Winding

Figure 5-13 is a simplified view using a lap winding, so-called because successive coils overlap each other. A partially developed view of this kind of winding is illustrated in Figure 5-14. The numbered rectangles at the bottom of the drawing represent the commutator segments, while the numbered vertical lines are the conductors embedded in the slots of the armature. The letters A and B indicate

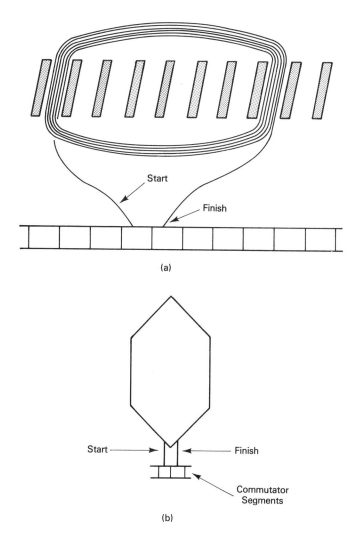

Figure 5-13 Lap winding (a); symbol (b).

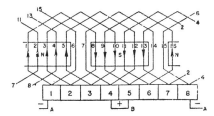

Figure 5-14 Partially developed view of a lap winding.

the brushes. If the start and finish leads of an armature coil do not cross, the connection is referred to as progressive. Figure 5-15 is a partial plan view and the symbol.

Referring again to Figure 5-14, the current from the source voltage flows from its negative (−) terminal, A, to commutator segment 1, through armature conductor 1, through conductor 8 to commutator segment 2, to conductor 3, to segment 3, through conductor 5, through conductor 12 and ending at the commutator segments connected to the positive (+) terminal of the voltage source.

The analysis of this winding can be simplified by writing it as

A-1-8-3-10-5-12-B   This is path 1. These numbers refer to conductors only.

A-4-13-2-11-16-B   This is path 2.

The plan view of Figure 5-14 can supply additional information, not immediately obvious. Thus, commutator segments 1 and 8 are connected, since both are negative voltage points. Segments 4 and 5 are also common connecting points for the positive voltage terminal.

Conductors 6 and 15 are thus joined to the negative terminal; conductors 7 and 14 are also joined at the positive terminal. Consequently there is a current from the minus terminal through coil conductors 6 and 15 and then through coil

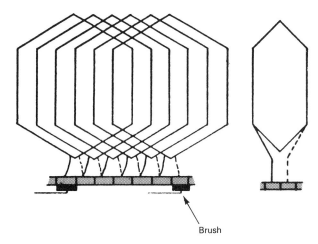

Brush

Figure 5-15 Plan view of a progressive winding and its symbol.

Types of Windings

conductors 7 and 14 to the plus terminal. The flow of current can be depicted as show by the arrows.

Progressive Winding

As indicated in the drawing of a simplex lap winding, the beginning and end leads of the armature coils are connected to adjacent commutator bars.

Coil Pitch

The span of a coil is the distance from one horizontal element of an armature coil to its other horizontal element. It is the number of slots between the positioning of the start and finish leads of an armature coil. If the sides of an armature coil are placed in slots number 1 and 18, the coil pitch is the difference between these two, or $18 - 1 = 17$. The coil pitch is 17. This doesn't necessarily mean that armature coil must start in slot number 1. It could start in 5, or 9, or 14, or any other slot. The selected slot is determined by the location of a pole.

Front Pitch and Back Pitch

The front of the armature coils is adjacent to the commutator; the opposite end is the rear. Both sides, front and rear, consist of the vertical lengths of the armature coil. The word *pitch* is also applied to these.

Pitch can be designated by the letter P. Front pitch could be identified as P_f and back pitch as P_b. More specifically, back pitch is the number of wire elements the coil advances on the back of the armature and is determined by the span of the horizontal elements. At the front of the armature the number of spanned elements is the front pitch, and this can be greater or less than the back pitch, but the two are never equal. In terms of formulas:

$$P_b = P_f + 2 \qquad P_b = P_f - 2$$

The formula with a plus sign indicates a progressive winding; that with a minus sign a retrogressive winding. A characteristic of the retrogressive winding is that the finish lead crosses the start lead. If a winding is changed from progressive to retrogressive, the effect will be to change the direction of rotation of the armature. The same effect will be noted in changing from retrogressive to progressive. The word *pitch* in general is used to indicate separation and can be applied to commutator segments, armature elements, and connecting leads from armature coils to commutator bars.

Commutator Pitch

While the words simplex, duplex, and triplex are used to indicate the separation between the start and finish leads of an armature coil, another method is to refer to such separation as *commutator pitch*. The separation or commutator pitch can be 14 if a coil connection is made to commutator bars 1 and 15.

Example:

What is the commutator pitch for a motor that uses 29 commutator segments and has four poles?

Solution: Four poles consist of two pairs of poles. Hence the commutator pitch in this case equals 29 plus 1 or 29 minus 1 divided by 2:

$$\frac{29 + 1}{2} = 15$$

$$\frac{29 - 1}{2} = 14$$

Commutator pitch, identified as P_c, determines the direction of rotation of the armature. It will rotate one way with a smaller C_p; the other way with a larger. Commutator pitch refers to the number of commutator bars separating the start and finish leads of a coil.

Retrogressive Winding

A retrogressive winding (Figure 5-16) is the opposite of a progressive type. In this case the end lead connection to a commutator bar precedes the bar connection of a beginning lead. If the front pitch is greater than the back pitch, the winding is retrogressive. This means the winding advances counterclockwise when viewed from the commutator end of the armature.

Simplex Winding

A simplex winding is one that has its armature coil beginning and end leads connected to adjacent commutator bars. In this description of an armature winding, the number of armature coil turns might not be listed, nor might any information

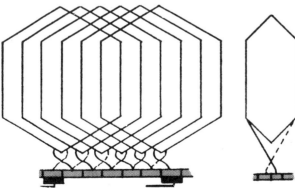

Figure 5-16 Plan view of a retrogressive winding and its symbol.

Types of Windings

be supplied as to how those turns are positioned in their slots. This is also applicable to other types of windings.

One of the advantages of the simplex winding is that it can be designed to have a higher current-carrying capacity. This larger current rating is achieved by having two or more coils wound in parallel.

Duplex

In a duplex arrangement the starting lead of an armature coil and the end lead are connected to commutator segments two commutator bars apart.

Triplex

As in the case of simplex and duplex, the word *triplex* refers to the separation of the connections of the start and finish leads of an armature coil. In a triplex winding these coil leads are connected three bars apart.

Simplex Progressive Lap Winding

Figure 5-17 shows the layout of a simplex progressive lap winding. A solid line is used to represent one coil; a dashed line another. As indicated in this illustration, there are two coils per slot. The two coils are in series, with their junction at the commutator slot centered between the beginning bar and the ending bar.

Figure 5-18 shows another drawing along the same lines and is also a simplex progressive lap winding, but this winding, as indicated, uses three coils per slot.

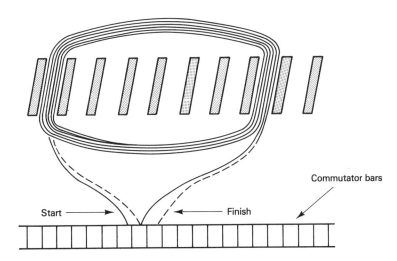

Figure 5-17 Simplex progressive lap winding of an armature. This arrangement uses two coils for each slot.

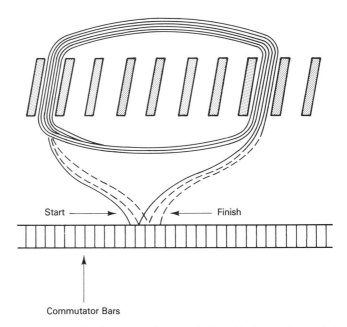

Figure 5-18 Simplex progressive lap winding with three coils per slot.

The three coils are in series. The start winding is connected to the first commutator bar, while the next two bars are junction points for the series coil. The last bar is used for the end lead of the three coils.

Total Lap Winding

The drawings showing the various lap windings are simplified types and are used to supply a general idea of the windings. A much more detailed illustration in Figure 5-19(a) is that of a complete winding. The motor's armature uses a duplex progressive lap winding having 24 slots, 24 commutator bars, and four poles. The coil span is $1 - 7$.

The center circle represents the 24 bars of the commutator.

Inside this circle are the four brushes. The outer circle depicts the slots—there are 24 of these. Each slot accommodates two horizontal elements of each coil. The solid lines represent that part of the coil positioned in the front part of the armature; the dashed lines, that part behind the armature. This isn't standard, though, since in some armature coil drawings the part of the coil in the front of the armature is drawn as a thick line; that part behind the armature as a thin line.

The active parts of each coil are the short horizontal lines positioned in the slots. This type of drawing can be best understood by tracing a single winding at a time.

Types of Windings

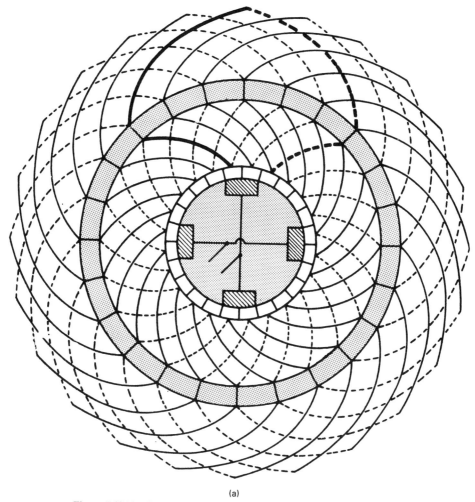

(a)

Figure 5-19(a) Duplex progressive lap winding. It has 24 slots, 24 commutator segments, 4 poles, and a coil span of 1 to 7.

Symmetrical and Non-Symmetrical Connections

There are several ways of connecting the beginning and ending of the armature elements to the commutator segments. One method is to make the connections as short as possible, a technique known as symmetrical. When the commutator bars are offset from the connecting wires, that is, separated from them by one or more bars, the connections are known as non-symmetrical (Figure 5-19[b]).

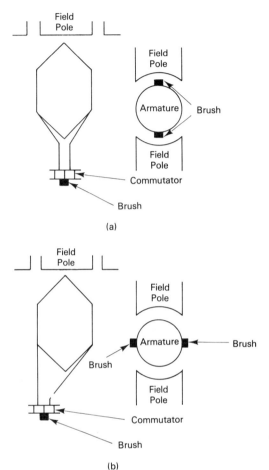

Figure 5-19(b) Symmetrical connections (a); non-symmetrical connections (b).

Wave Windings

As in the case of lap windings, wave windings can be connected in simplex, duplex, and triplex, and can also be progressive or retrogressive. The name *wave winding* is obtained from the way in which the current circulates through the armature. The basic difference between a lap winding and a wave winding is in the way the coil connections are made to the armature (Figure 5-19[c]).

ARMATURE POLES

When an electrical current supplied by the source voltage flows through the armature coils, these each become electromagnets having north and south poles. There are two poles for each coil, whether the coils are connected in shunt or

Armature Poles

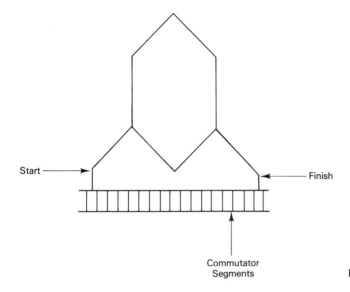

Figure 5-19(c) Wave winding.

series. The minimum number of poles is two, assuming just a single coil is energized. The number of poles is used as one way of identifying the characteristics of a motor: a two-pole, four-pole, six-pole, and so on.

Force Exerted on Armature Elements

Because of the presence of a magnetic field around the poles of the field magnets, a force is exerted on the conductive horizontal elements of each armature coil. The force on the elements, while the elements are in the presence of this magnetic field, is directly proportional to its strength, the length of the conductive element, and the amount of current flowing through it. The force can be calculated (in the English system) by

$$F = \frac{8.85\ BLI}{10^8}$$

where

F = the force exerted on the armature element in pounds,

B = the flux density in lines per square inch,

I = the current in amperes, and

L = the length of the conductive element of the armature coil in inches.

This applies to each element of the armature coil. If a single coil is used, there are two such elements, and the result of the formula should be multiplied by 2. If a coil has six elements multiply by 6.

Example:

The active length of a single element of a two-turn armature coil is 1 ft. The current flow through this element is 8 A. The flux density is 60,000 lines per square inch. What is the force in pounds exerted on the element?

Solution:

$$F = \frac{8.85 \; BLI}{10^8}$$

$$= \frac{8.85 \times 60,000 \times 12 \times 8}{10^8}$$

$$= \frac{8.85 \times 6 \times 10^4 \times 96}{10^8}$$

$$= \frac{8.85 \times 6 \times 96}{10^4} = \frac{5097.6}{10,000} = 0.5098 \; \text{lb}$$

If this is the upward force exerted on one element of the armature coil, then a similar downward force is exerted on the second element of the coil. The total force is then $2 \times 0.5098 = 1.02$ lb approximately.

When the flux lines of a magnetic field are at right angles to the conductor of an armature coil, the force that is exerted on the conductor can also be given by

$$F = \frac{BLI \times 10^{-7}}{1.13}$$

When this formula is applied to the preceding problem:

Solution:

$$F = \frac{60,000 \times 12 \times 8 \times 10^{-7}}{1.13}$$

$$= \frac{6 \times 12 \times 8 \times 10^{-3}}{1.13} = \frac{6 \times 12 \times 8}{1,130} = 0.5097 \; \text{lb}$$

The force exerted on a conductor in a magnetic field can also be expressed (in the MKS system) as

$$F = BLI$$

where

F = the force in newtons,

B = the magnetic density in webers per square meter,

L = the length of the conductor in meters, and

I = the current in amperes.

Field Coils

The resulting torque is the product of this force and the radius of the rotating armature.

Eddy Current Power Loss in Laminated Cores

Even though the cores for armature and field coils are laminated and are insulated from each other by natural oxides that are deliberately allowed to accumulate on their surfaces, or are coated with an insulating varnish, there is still some power loss. This loss expresses itself in the form of heat. The power loss, in watts per cubic centimeter, is

$$W = \frac{1.64 t^2 f^2 B^2}{10^{16} \rho}$$

where

t = the thickness of the laminations in centimeters,

f = the frequency in hertz (cycles per second),

B = the flux density in gauss, and

ρ = the resistivity of the material in ohms per cubic centimeter.

The formula indicates that the power loss, in watts, varies as the square of the lamination thickness, the square of the frequency, and the square of the flux density. To minimize this loss it would be desirable to keep the laminations as thin as possible.

Eddy current loss can also be calculated from

$$W = k t^2 f^2 B^2$$

where

W = the power loss in watts,

k = a constant for a given specimen,

t = the thickness of lamination in meters,

f = the frequency in hertz, and

B = the flux density in tesla.

FIELD COILS

For each pair of armature poles there is a corresponding pair of field poles. If a motor is identified as a four-pole type, this infers there are four field poles.

Also known as the *stator* when motionless, the field coils can use the same source voltage as the armature, and this can be a dry cell, a wet or gel-type cell, a DC power line, or an electronic power supply.

The field coils, when current-operated, are electromagnets. Since they do not rotate, connections to the source voltage are simple and direct. Field coils, like the armature, are wound on laminated iron and are fastened to the inner part of the motor yoke or frame.

The energy in a magnetic field, whether it surrounds the field coils or the armature, can be expressed by

$$W = \tfrac{1}{2}LI^2$$

where

W = the energy in joules or watt-seconds,
L = the inductance of the coils in henrys, and
I = the direct current in amperes.

INTERPOLES

One of the problems of using carbon brushes as a means of bringing an electrical current into a rotating armature is the sparking that can occur between those brushes and the commutator bars. This sparking can be minimized by using a small auxiliary pole, supplied by an electromagnet and positioned between the main field-coil poles. The supplementary magnets are also known as *commutating poles.*

Another technique for reducing commutator sparking is to shift the brushes, but this requires a brush holder that can be moved left or right around the circumference of the commutator.

MOTOR FRAMES

The motor frame, also called an *enclosure* or a *yoke,* is the housing for the motor parts. It has both mechanical and electrical characteristics. Mechanically, it guards the motor against dust, atmospheric corrosion, and water. It acts as a heatsink, transferring heat developed by the motor to the air, thus preventing excessive heat buildup inside the motor. Its base is used as a motor support. Electrically, it minimizes the radiation of electrical interference. Because of its ferrous structure it adds to the inductance of the field coils, which are mounted on its interior surface. Not all motor frames are alike, with each designed for a specific function. The various types of frames used for electric motors are listed in Table 5-4.

Open Frame

The open frame is characterized by large openings that permit the escape of heat. This design is intended for motors that develop quantities of heat which cannot be completely radiated by an all-enclosing frame. An external blower may be part of

Motor Frames

TABLE 5-4. MOTOR FRAMES

Open
Totally Enclosed
Protected
Drip Proof
Splash Proof
Dust Proof
Watertight
Explosion Proof
Ribbed
Vented
Internal Fan Cooled
Heat Sink
Totally Enclosed, Fan Cooled
Open
Conduit Box Equipped

the motor frame, with the blower force feeding outside air to all operating parts. The open-frame design is often used when motors must be operated continuously.

Totally Enclosed Frame

This frame encloses the motor completely and is required when the motor is to work in a hostile environment. It is commonly found in small and miniature motors, too, as a means of protecting fingers from any rotating parts. In large motors it may be used as a shield against dust or chemical fumes. The large sizes may be ventilated by a blower.

Protected Frame

Somewhat similar to the totally enclosed frame, this type has air vents covered with screening. Also known as semi-enclosed, the screening not only permits greater air circulation but works to prevent accidental finger contact with the rotor.

Drip-Proof Frame

This frame supplies motor protection against liquid drip but with provision for ventilation. The assumption is made that any possible drip falling on the motor at any angle not greater than 15° off the vertical cannot enter.

Splash-Proof Frame

This type of motor enclosure is designed so that liquids or solid particles falling vertically or reaching the frame at some angle not greater than 100° from the vertical cannot reach the interior of the motor.

Dust-Proof Frame

The motor frame is designed so the motor is completely enclosed, not permitting dust entry from any angle.

Watertight Frame

With this frame the motor is enclosed so completely that water applied from a hose, at any angle, cannot reach the motor interior.

Explosion-Proof Frame

This frame is a variation of the totally enclosed type and is designed to withstand an internal motor explosion. The purpose is to prevent any ignition of gas or dust in the motor's environment.

Frame Variations

Heat dissipation. The exterior of the motor frame may be ribbed when that frame works as a heat sink, so as to provide a greater heat radiating surface.

Vented. Vents may be provided on the front of the motor near the exit portion for the motor shaft.

Internal fan cooled. These motors have vents around the edges of the frame.

Cooling Bands. This is a completely shielded frame equipped with a heat sink enclosing the motor on the outside circular part of the frame. It may consist of a ribbed structure or metal strips in the shape of bands.

Totally enclosed, fan cooled. This motor has a circular frame, but the frame is separated a short distance from its circular end plates.

Open construction. This motor is completely exposed and has no enclosure.

Motor frame with conduit box. This frame is designed to accommodate a conduit box for electrical connections. The box is completely enclosed to prevent accidental touching of power-line connections.

Totally enclosed. While this motor is enclosed it has provisions for side-mounted electronics.

Mounts. The motor frame may have a rigid base, mounting lugs, a resilient mount, weld bolts, belly bands, face mounting nuts, extruded and tapped holes, and through-bolts.

MOTOR BEARINGS

The motor shaft rests on end bells that support bearings. The bearings (Figure 5-20) not only support the shaft but do so with minimum friction. Bearing types include ball, roller, and sleeve.

The difference between the outside diameter of the shaft and the inside diameter of the bearing is between 0.00020 and 0.0010 in. The shaft clearance ranges from 0.00010 to 0.00050 in. These dimensions do not apply to all motors but indicate tolerance tightness. The shaft and all the components mounted on it or pressing against it must be supported by the film of oil in the space between the shaft and its bearings.

Bearing Oil

Bearing oil must be the type recommended by the motor manufacturer. Sleeve bearings use pure mineral oil only. Factors determining the motor oil to use are listed in Table 5-5.

MOTOR SHAFT

Made of hardened steel, the motor shaft is available in various lengths, diameters, and shapes. Three of the more commonly used are illustrated in Figure 5-21. That in drawing (a) is a round type, (b) is a flatted round, and (c) is a knurled round type.

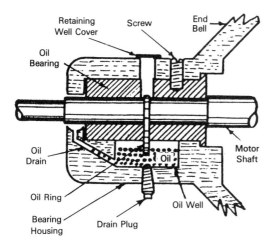

Figure 5-20 Cross section of bearing.

**TABLE 5-5. FACTORS
DETERMINING MOTOR OIL
SELECTION**

Motor operating temperature
Ambient temperature
Viscosity
Rust prevention
Wear ability
Cooling ability
Temperature coefficient of expansion

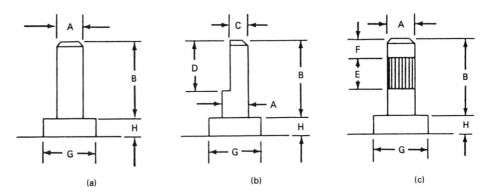

Figure 5-21 Representative motor shafts: (a) round; (b) flatted round; (c) knurled round. (Courtesy Cramer Co.)

The letters adjacent to the shafts in these drawings indicate dimensional specifications: A is the shaft diameter, B the free shaft length, C the width across the top of the shaft, D the flat length, E the length of the knurled portion of those shafts so manufactured, F the knurl setback, G the bearing outside diameter, and H the bearing height.

THE SERIES-WOUND MOTOR

Figure 5-22 shows the diagram for a series wound motor. It is so called since the field and armature windings are in series with each other, thus the same current level flows through both. Table 5-6 lists the characteristics of this motor.

Speed Control

The speed can be controlled by connecting a variable wire-wound resistor, actually a rheostat, in series with the field coil as shown in Figure 5-22(b). The resistor has a relatively low value of resistance.

The Series-Wound Motor

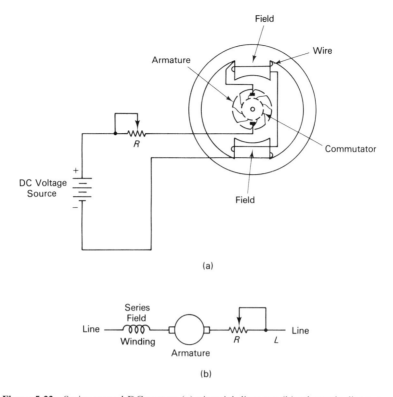

Figure 5-22 Series-wound DC motor: (a) pictorial diagram; (b) schematic diagram.

TABLE 5-6. CHARACTERISTICS OF THE SERIES-WOUND DC MOTOR

1. Used mainly for fluctuating loads. Motor can start with heavy overload.
2. Poor speed regulation. The greater the load, the slower the speed. Conversely, the lighter the load, the greater the speed. Suitable for repeated acceleration with heavy loads. Same current flows through field coils and armature.
3. Current flow can increase sharply.
4. Motor races when load is removed.
5. Torque varies approximately as the square of the field or armature current. Torque must not go below 10% to 15% of full-load torque.
6. Most DC motors are equipped with interpoles. Interpoles are wired in series with field and armature coils.
7. Direction of rotation can be obtained by reversing direction of current flow.
8. Field is constructed of a few turns of heavy wire, that is, wire having a fairly low gauge number.
9. Field strength, under normal operating conditions, varies with the armature current.
10. Excellent starting and stalling torque.

The speed of a series wound DC motor can be calculated in several ways. One formula is

$$S = K \left[\frac{V_a - I_a (R_a + R_e) - E_b}{I_b} \right]$$

where

S = the speed of rotation in revolutions per minute,
V_a = the voltage applied across the motor,
I_a = the motor current in amperes,
I_b = the brush current in amperes,
R_a = the resistance of the series field in ohms,
R_e = the resistance of the armature in ohms,
E_b = the voltage drop across the brushes, and
K = a constant for a particular machine = $I_a/k\phi$, in which
ϕ = the flux.

Another formula arrangement is

$$S = \frac{V_a - I_a (R_a + R_e) - E_b}{k\phi}$$

For both formulas, the expression contained in the numerator is identical.

SPLIT-FIELD SERIES-WOUND MOTOR

The split-field series-wound DC motor is a variation of the single series-field type. It is known as a split-field because it has a tap at the electrical center of the field coil, as shown in Figure 5-23. This tap, plus a single-pole double-throw (SPDT) switch permits an easy change of armature rotation. The switch changes the direction of current flow from the voltage source, transposing the polarity of the field electromagnet. All the characteristics of the series-wound DC motor listed in Table 5-6 apply.

SHUNT-WOUND MOTOR

In the shunt-wound motor (Figure 5-24) the armature and field coils are connected in parallel (shunt), with both wired across the DC source voltage. The field coil is wound with many turns of fine wire. Although the field current is much less than

Shunt-Wound Motor

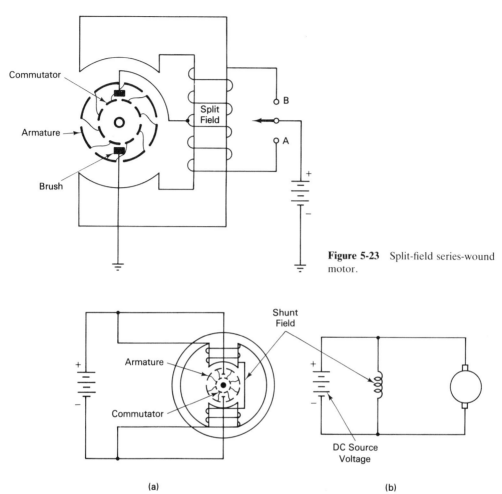

Figure 5-23 Split-field series-wound motor.

Figure 5-24 Shunt-wound motor: (a) pictorial; (b) circuit diagram.

that used in a series wound, the ampere-turns (NI) is substantial. The resistance of the armature is very low and can be less than 1 Ω. The characteristics of the shunt-wound DC motor are detailed in Table 5-7.

In a shunt-wound DC motor the total current from the source is equal to the sum of the field and armature currents. That is,

$$I_{line} = I_{field} + I_{armature}$$

The current flowing through each of these components, the armature and the field, is inversely proportional to their respective resistances. However, since they are in shunt, the same voltage appears across each. Expressed as a formula,

$$I_1 \times R_1 = I_2 \times R_2$$

TABLE 5-7. CHARACTERISTICS OF THE SHUNT-WOUND DC MOTOR

1. Practically constant speed for varying load conditions.
2. At full speed without a load counter EMF is almost equal to the source voltage.
3. Speed can be controlled by variable resistor in series with the field or by adjustment of the source voltage.
4. Starting torque is less than that of the series-wound motor.
5. If shunt field is open, armature speed rises substantially.
6. When motor is started, field winding should receive excitation current first.

where

I_1 = the armature current,

R_1 = the armature resistance,

I_2 = the field current, and

R_2 = the field resistance.

This formula permits determination of any one of these quantities, provided the values of the remaining three are known. Thus

$$I_1 = \frac{I_2 \times R_2}{R_1}$$

$$I_2 = \frac{I_1 \times R_1}{R_2}$$

$$R_1 = \frac{I_2 \times R_2}{I_1}$$

$$R_2 = \frac{I_1 \times R_1}{I_2}$$

FIELD CURRENT IN A SHUNT MOTOR

The field current can be calculated by

$$i = \frac{V}{R_t}\left(1 - e^{-\frac{1}{R_f L_f}}\right)$$

where

i = the DC field current at time t after energization,

V = the DC voltage applied to the field,

R_f = the resistance of the field in ohms,

Motor Operating Voltage

L_f = the inductance of the field in henrys,

t = the time after the field is energized in seconds, and

e = 2.718 (the base of natural logarithms).

MOTOR OPERATING VOLTAGE

Every motor works not only as a motor but as a generator, and in so doing develops a counter EMF (CEMF) The CEMF must always be a little less than the applied line voltage, or else no current would flow into the motor. The actual operating voltage of the motor is

$$E = E_{line} - CEMF$$

where

E = the motor operating voltage,

E_{line} = the source voltage, and

$CEMF$ = the voltage generated by the motor.

Example:

A shunt wound DC motor is connected to a 121-V DC generator. The power used by the motor is 18.7 kW (18,700 W). Power losses in the armature in the form of heat are 395 W. The resistance of the field windings totals 92.8 Ω. Determine the amount of line current, the field current, the resistance of the armature winding, and the CEMF.

Solution: One of the power formulas is

$$P = E \times I$$

The line current can be calculated from

$$I_L = P_t/E_L$$

where

I_L = the line current,

P_t = the amount of power taken from the voltage source, and

E_L = the line voltage.

Thus,

$$I_L = 18.7 \text{ kW}/121 = 18,700/121 = 154.55 \text{ A}$$

The field current can be calculated from

$$I_f = E_L/R_f$$

where

I_f = the field current in amperes,
E_L = the line voltage, and
R_f = the resistance of the field winding.

Thus,

$$I_f = 121/92.8 = 1.3 \text{ A}$$

The armature current is

$$I_a = I_L - I_f$$

where

I_a = the armature current in amperes,
I_L = the line current in amperes, and
I_f = the field current in amperes.

Thus,

$$I_a = 154.55 - 1.3 = 153.25 \text{ A}$$

One of the basic formulas for calculation power is $P = I^2R$. The armature resistance can be determined by using this formula with a transposition, that is, $R = P/I^2$. Thus the armature resistance is

$$R_a = P_a/I_a^2$$
$$R_a = 395/153.25^2 = 0.01682 \text{ }\Omega$$

The counter EMF can be calculated from

$$CEMF = E_L - I_aR_a$$
$$CEMF = 121 - (153.25 \times 0.01682)$$
$$= 121 - 2.577665$$
$$= 118.422 \text{ V}$$

COMPOUND-WOUND MOTOR

The compound-wound DC motor has some of the characteristics of the series-wound and shunt-wound units. The compound-wound's characteristics are listed in Table 5-8.

Figure 5-25 illustrates the pictorial and the circuit diagram for this motor.

Differentially Wound Compound Motor

TABLE 5-8. CHARACTERISTICS OF THE COMPOUND-WOUND MOTOR

1. Large starting torque.
2. Moderately constant speed under fixed load or for quick changes from light to heavy loads. The speed does not change as much as a series-wound type but more so than a shunt wound.
3. No-load speed is low.
4. No danger of armature runaway with light load or no load.
5. Equipped with both a series and shunt field windings.
6. Shunt and compound motors are the more widely used DC types.
7. Compound wound are generally used for larger motor types.
8. Speed controllable by use of series and shunt resistances.
9. Reverse rotation can be had by interchanging leads at brush holders. Motor may be equipped with a reversing switch.

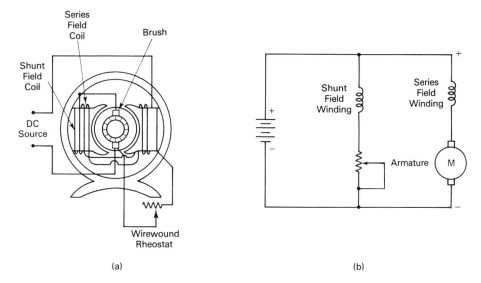

Figure 5-25 Compound-wound motor: pictorial (a); schematic (b).

DIFFERENTIALLY WOUND COMPOUND MOTOR

The differentially wound compound motor is a member of the compound motor family. Usually, in a compound-wound machine the magnetic field surrounding the shunt winding aids that around the series field. In the differentially wound motor (sometimes called a *flat-compounded motor*) the two fields, series and shunt, are wound so that their magnetic fields are in opposition. The advantage of this arrangement (Figure 5-26[a]) is that it supplies automatic regulation of the armature speed under varying load conditions.

Assume the load on the machine is such that the armature is rotating at its normal speed. If the load is increased, the speed of the armature will slow slightly.

This will result in a smaller value of counter EMF, allowing more current to flow from the source voltage. But the increased current through the armature will also flow through the series field coil, thus increasing its magnetic field intensity. This larger magnetic field strength of the series coil will oppose the magnetic field of the shunt coil. This action permits the armature to increase its speed.

If, on the contrary, the load on the machine is decreased, exactly the opposite effect takes place. The flow of direct current through the armature and the series coil is reduced. This permits a greater current flow through the shunt field, increasing its magnetic strength, preventing the armature from excessive speed rotation.

CUMULATIVE-WOUND COMPOUND MOTOR

The differentially wound and the cumulative wound are two types of compound-wound motors. The cumulative wound (Figure 5-26[b]) has the high starting torque of the series-wound motor but because of the presence of a shunt field does not have its "runaway" characteristic. Unlike the differentially wound motor, in which the current through the series field and shunt field moves in opposite directions (thus producing opposing magnetic fields), the flow of current in the cumulative-wound compound motor through the series and shunt fields is in the same direction. This results in magnetic fields that aid each other.

The circuit diagrams in Figure 5-26 are those of the differentially wound and cumulative-wound motors.

Circuitwise the two are identical except for the direction of current flow through the series field winding. The direction of the current flow is determined

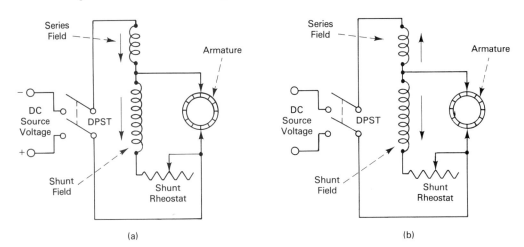

Figure 5-26 Comparison of two types of compound-wound motors: differentially wound (a); cumulative wound (b). Arrows indicate direction of current flow. DPST is a double-pole, single-throw switch.

Cumulative-Wound Compound Motor

by the way in which the series field coil is arranged physically. Here the direction of current flow is controlled by the way the series field coil is positioned with respect to the shunt field coil.

Compound-wound motors do not have the heavy-load starting capabilities of the series-wound DC motors but are better in this respect than shunt-wound types.

The speed (S) in rpm of any DC differential or any cumulative-wound compound motor can be calculated from

$$S = \frac{V_a - I_a(R_a + R_s) - E_b}{k(\phi_f - \phi_s)}$$

where

R_s = the resistance of the series field in ohms,

ϕ_f = the shunt field flux,

ϕ_s = the series field flux,

V_a = the armature voltage,

I_a = the armature current,

R_a = the armature resistance,

k = a constant (see equation below), and

E_b = the voltage drop across the brushes.

The general equation for the speed of any DC motor is

$$S = \frac{V_a - (I_a R_a + E_b)}{k\Phi}$$

where

S = the speed of rotation in rpm,

V_a = the voltage applied to the armature terminals,

I_a = the armature current in amperes,

R_a = the armature resistance in ohms,

E_b = the voltage drop across the brushes,

ϕ = the flux lines per pole linking the armature conductors, and

k = the constant for any given machine =

$$\frac{ZP \times 10^{-8}}{60a}$$

where

P = the number of poles,

Z = the number of active conductors on the surface of the armature, and

a = the number of parallel current paths in the armature winding.

The current taken by a DC motor can be calculated by

$$I = \frac{hp \times 746}{E \times eff}$$

where

hp = the horsepower output,

I = the current in amperes,

E = the DC source voltage, and

eff = the efficiency stated as a decimal.

Note: As a general rule for 230-V DC motors the current is approximately 4 A per horsepower.

As an alternative to the use of a formula, Table 5-9 supplies the current requirements of motors having various horsepower requirements. The values supplied are for full load. Do not consider the data as precise, since there can be a tolerance factor of $\pm 10\%$.

The wire size for DC motors is also supplied in Table 5-9.

TRANSMISSION OF POWER

The transmission of power by the motor shaft can be calculated from

$$hp = \frac{torque \times rpm}{5{,}250}$$

where torque is in foot-pounds.

TABLE 5-9. WIRE SIZE FOR DC MOTORS

hp	Approx. full load amps	Copper wire size min. AWG	Branch current fuse-amps
1	4.8	14	15
2	8.5	12	15
3	25	8	40
5	40	6	60
10	76	2	125
15	112	00	175
20	148	4/0	225

NAME PLATE (DATA PLATE)

Most motors bear a data plate attached by the manufacturer. The data plate may include such information as is listed in Table 5-10. The amount of data supplied can vary from manufacturer to manufacturer and is not standardized.

STEPLESS MOTORS

The shaft of a motor rotates, sweeping through a succession of complete circles, each of which is 360°. Because of the repetitive nature of the turning of the motor shaft, such a motor is referred to as *stepless*. If the shaft can be made to turn in the opposite direction as well it is still a stepless type, but to differentiate it from motors that turn in one direction only can be categorized as *bidirectional stepless*. In a stepless motor, whether unidirectional or bidirectional, there is no control of the amount of angular displacement of the motor shaft.

STEPPING MOTORS

Unlike stepless motors, the stepping type (also known as a *stepper*) is one in which the shaft of the motor can be made to rotate a precise number of degrees, ranging from one or more complete turns, to some fraction of a turn. The rotation is measured in degrees, and since a single complete turn is 360° such motors are characterized by some specific fraction of a whole turn and are listed by steps. These steps can be a selected portion of 360°, such as 20°, 30°, 60°, and so on. Each selected step can be repetitive and so can the time interval between steps. Stepping motors are available in a number of different variations in construction.

Types of Stepping Motors

There are various types of stepping motors, as indicated in Table 5-11.

TABLE 5-10. TYPICAL DATA PLATE INFORMATION

Model number
Serial number
Current
Designation of DC or AC
Frequency
Line voltage

TABLE 5-11. TYPES OF STEPPING MOTORS

Disk stepper
Variable reluctance
Permanent magnet
Hybrid

Steps versus Step Angle

The larger the number of steps, the smaller the step angle. Step angle can be obtained from:

$$\text{Step Angle} = 360/\text{number of steps}$$

Step angle is also known as *step size* and is always in degrees. The step angle can be a whole number plus a decimal fraction, provided it is divisible into 360°. Typical stepping angles are 7.5°, 9°, 11.25°, 15°, and 18°. Thus, $360/7.5 = 48$ steps; $360/9 = 40$ steps; $360/11.25 = 32$ steps.

Step Frequency

Stepping time is also a consideration. Steps per single complete revolution of the motor shaft are commonly 24, 32, and 48 steps per second.

Microsteps

A full series of steps, such as 24, 32, and 48, can be subdivided into much smaller steps, controlled by the drive signal pulse frequency.

Incremental Mode

Incremental mode means that the motion of the rotor shaft isn't a continuous, unbroken revolution, but proceeds in specific amounts or increments. If a rotor turns 10° per input signal pulse, it will complete 36 increments in one complete revolution of its shaft.

Disk stepper. The disk stepping motor, also called a *disk stepper,* is illustrated in Figure 5-27. It consists of a samarium-cobalt disk centered and mounted on a shaft. The disk is free to rotate between a pair of stator poles, shown in the drawing as phase A and phase B, though they are sometimes referred to as *phase 1* and *phase 2*. These alternative names emphasize the fact that it is a two-phase motor. Each phase is electrically shifted by 90° with each phase consisting of one or more windings. The phases can be wired in parallel, as in Figure 5-28(a), or in series, as in Figure 5-28(b). The number of phases is independent of the actual number of coils per phase, most often four. The driving or excitation voltage for

Stepping Motors

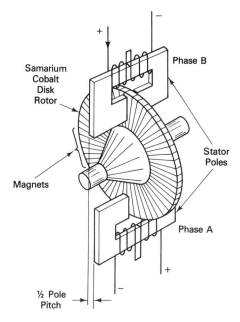

Figure 5-27 Basic stepper motor. (Courtesy Penton/PC, Inc.)

the coils consists of pulses. Step action is determined by pulse frequency. Half stepping can also be obtained.

The drawing in Figure 5-29 has 50 magnets, although a stepper motor can have more or fewer. Those in this illustration have magnets that are 0.7 mm long, with all having the same thickness as the disk. The larger the number of magnets the smaller the step angles that can be obtained, thus increasing the total number of available steps. The cross section of each magnet is approximately 1.5 × 1.5 mm. The diameter of the disk can range from 30 mm to 100 mm.

Variable reluctance (VR). Variable reluctance motors do not use permanent magnets (Figure 5-30). They have a soft-iron multiple rotor and a two-to-five-phase stator. The flux path originates at the energized stator pole, goes through the rotor when the teeth are aligned, attracts the nearest rotor tooth, and aligns it with the energized stator tooth. The number of teeth on the rotor and stator as well as the number of phases determines the step angle. The stator teeth are wound with wire. A group of windings forms a phase.

A step takes place when one phase is de-energized and the next phase in sequence is energized. The rotor will then move to a new position of minimum reluctance with the rotor and stator teeth lined up. This completes one step.

Permanent-magnet (PM) rotor. The permanent-magnet-rotor stepping motor is also known as a *stamped-construction* or *sheet-metal stepper motor*. The rotor is radially magnetized, with alternate poles. The stator (Figure 5-31) has two halves, each of which contains a coil. The halves have teeth to direct the magnetic

202   DC Motors   Chap. 5

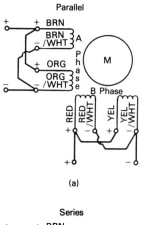

(a)

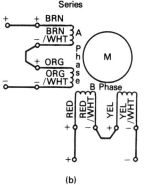

(b)

Figure 5-28 Phasing coils wired in parallel (a); in series (b). (Courtesy Portescap)

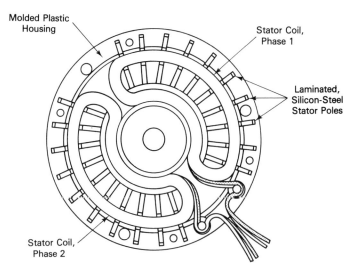

Figure 5-29 Face view of stator of disk stepper. (Courtesy Penton/PC, Inc.)

Stepping Motors

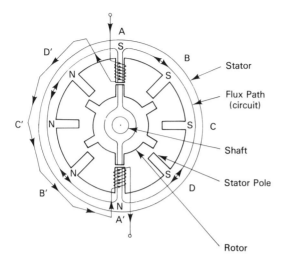

Figure 5-30 Variable-reluctance stepping motor. (Courtesy Small Motor Manufacturers Assn.)

flux path in and out of the PM rotor. This motor is economically competitive, but its performance is not as good as other steppers in terms of accuracy and speed. Because it uses a permanent magnet, this motor requires less operating power than other types. It also has better damping, that is, better settling to a dead stop. It is not suited for small step angles.

Hybrid. Also called a 1.8° PM stepping motor, the hybrid uses the stator of a variable-reluctance stepping motor and a permanent magnet in the rotor. The phase windings are excited with DC pulses. The motor has a high efficiency, is capable of high step rates, and is available in step angles of 1.8°.

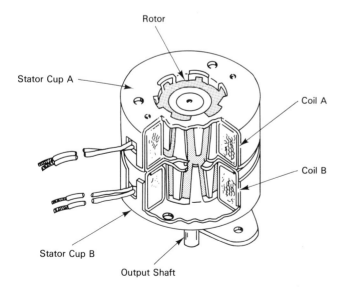

Figure 5-31 Permanent-magnet stepping motor using cylindrical magnet design. (Courtesy Small Motor Manufacturers Assn.)

The rotor has three main parts: a cylindrical permanent magnet and two rotor sections on each end. The stator has salient poles (multiple teeth on one pole) and usually bifilar-wound coils.

Terminology

The terminology of stepper motors is listed in Table 5-12.

INSULATION

Insulation has several purposes. Primarily it is intended to keep conductors from shorting, but it also protects those conductors from environmental hazards, including both gases and liquids. A disadvantage is that insulation restricts the heat-radiating ability of the conductors.

Table 5-13 is a listing of the commonly used insulating materials.

While insulation is commonly thought of with respect to conductors, it is also required for motor accessories, such as switches, control electronics, transformers, and so on. In the case of large transformers, an insulation oil is used to conduct heat away from the windings to the transformer case, whose ribbed structure works as a heat sink.

TABLE 5-12. TERMINOLOGY OF STEPPER MOTORS

Bipolar Drive. A drive circuit that creates an excitation causing torque and producing reversals in the windings.
Chopper Drive. A drive circuit with high voltage applied to the phase windings until the winding current reaches a predetermined value, at which time the supply is switched off. When the current decreases to a second set value, the supply is switched on.
Constant Current Drive. A drive circuit that maintains nearly constant current in the stator windings at any rotor speed.
Damping. Elimination or reduction of step overshoot. Includes mechanical, electrical and viscous types.
Drive. Circuitry that controls the stepping motor.
Half Step. One-half specified step angle.
Holding Torque. Amount of torque required to break the shaft away from its holding position.
Load Angle. Angle between magnetic axis of stator and rotor.
Pull-In Rate. Maximum pulse rate at which stepping motor will start.
Pull-In Torque. Minimum torque at which stepping motor will start.
Pull-Out Rate. Constant maximum pulse rate at which motor will run.
Pull-Out Torque. Maximum torque applied to shaft of the motor.
Ramping. Controlling the rate of change of the pulse frequency so as to accelerate or decelerate the motor.
Residual or Detent Torque. Amount of torque required to break the shaft away from its holding position.
Resonance. Natural frequency characteristics of the motor.
Step. Rotation of the motor shaft by the step angle; z is the letter symbol.
Steps per Revolution. Steps required for one complete revolution of the shaft.

TABLE 5-13. INSULATING MATERIALS

Natural and synthetic rubber
Cotton
Silk
Varnished cambric
Paper
Asbestos
Slate
Mica
Porcelain
Glass
Plastics
Tape
Laminates
Films and sheets
Shellac
Varnish
Enamel
Oil

MOTOR SYMBOLS

Various symbols are used to indicate motors in circuit diagrams. These symbols are illustrated in Figure 5-32. There is no standardization, so any one of these, or possibly others, are used. The symbol gives no indication of the motor type, but this data is sometimes printed alongside the symbol.

BRUSHLESS DC MOTORS

These are permanent-magnet motors using solid-state electronic switches to send power to their windings. They do not use brushes or a commutator. The windings are placed in the stator, and permanent magnets are attached to the rotor (Figure 5-33).

Brushless DC (BLDC) motors have a wide speed range, sometimes in excess of 50,000 rpm. Operating speed is dependent on the load (see Table 5-14). The source voltage is low and is typically less than 42 V DC, but there are lower and higher voltages, 24 V being common.

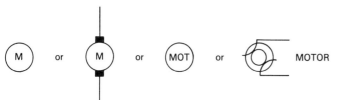

Figure 5-32 Generalized motor symbols.

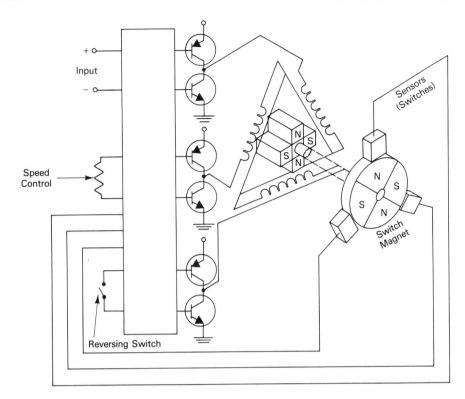

Figure 5-33 Operational setup of brushless DC motor. (Courtesy Eastern Air Devices, Inc.)

TABLE 5-14. OPERATING CHARACTERISTICS OF SOME SELECTED BRUSHLESS DC MOTORS

Drive* circuit	Full-load amperes	Full-load torque	Full-load rpm	No-load rpm	Volts	Starting torque (ounce-inches)
B	3.0	8	6,000	8,500	24	48
B	4.5	30	3,000	5,000	24	80
B	4.0	10	10,000	12,500	24	60
U	0.5	4	2,200	3,000	24	14
U	1.2	4	4,400	6,000	24	8
B	8.0	38	5,000	5,600	24	112
B	5.0	28	4,000	4,800	24	125
B	10.0	56	4,000	5,300	24	200

*U = Unipolar; B = Bipolar.

Brushless DC Motors

Applications

BLDC motors, also known as *electronically commutated motors,* are used in office equipment, copying machines, fans, blowers, pumps, centrifuges, scanners, tape drives, and so on.

Operational Setup

Hall-effect devices near the rotor sense its position and supply this information to a drive chip. This chip activates the power transistors in the proper sequence so as to supply power to the windings, creating a rotating magnetic field. The electronic commutator switches the stator windings at the appropriate rotor positions to generate continuous rotation and torque.

BLDCs use outer stator laminations with rotating permanent magnets (Figure 5-34). The rotor magnets are ferrite, rare-earth, or Alnico. Hall-effect devices, optics, or vane-coupled coils are used for commutation. These motors can be built with the rotor inside and this is standard, or outside, known as inverted construction.

There are two types of speed-controls: velocity feedback and phase-locked loop.

Slotless Stator

With high-energy magnet materials it is possible to make a surface-wound armature without slots. With a brushless DC motor the field generally rotates and the armature is stationary. Hall-effect devices are sensitive to magnetic flux density and supply a voltage of about 5V when they see a north pole, and 0 V for a south pole, supplying an analog signal roughly proportional to flux density.

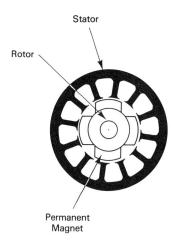

Figure 5-34 Motor construction. (Courtesy Eastern Air Devices, Inc.)

TABLE 5-15. ADVANTAGES AND DISADVANTAGES OF OUTER AND INNER ROTORS

	Outer rotor	Inner rotor
Inertia	Higher—sometimes an advantage	Lower—sometimes an advantage
Torque/Power	Higher	Lower
Cost	Lower	Higher
Balance	More critical	Less likely to need balancing.
High speed	Magnets on inside diameter. Bonding not critical.	Magnets on outside diameter.
Printed circuit board	Usually easier to package Hall devices.	Hall-device packaging can be a problem on small motors.

Using an inner rotor can have its disadvantages as well as advantages, as indicated in Table 5-15.

Miscellaneous Data

BLDCs are made in different frame sizes, ranging from 1 to 5.6 in. The frame can be ventilated or enclosed. Output power is as high as 1/2 hp, but some models are available with 1/100 to 1/5 hp for continuous duty. Motor efficiencies range from 30% to 85%. Rotation can be clockwise, counterclockwise, or reversible.

chapter six

AC Motors

AC motors are used extensively in the home and industry, and in transportation where an AC source is available either directly through the use of an independent generator, or indirectly from a motor generator. AC motors are available in a large variety (Figure 6-1), but quite often a so-called new type is simply a modification of some already existing motor. Table 6-1 is a listing of AC motors.

It is also possible for the same AC motor to be available under different names. A listing of AC motors could include AC series, autotransformer capacitor, hybrid, motors with built-in electronics, motors using external electronics, multi-speed, dual voltage, geared, subfractional horsepower, unidirectional, bidirectional, induction hysteresis, and so on. Motors can also be identified as single-phase or polyphase. Single-phase motors include repulsion, induction, series, capacitor, and synchronous. There are also some variations under these general headings.

BASIC AC MOTORS

There are two basic differences between AC and DC motors. The first is that the driving source voltage is AC instead of DC. The second is that because the input is AC, a commutator is usually not required. However, some AC motors do use commutators, as in the case of universal and repulsion types.

Since the armature is a rotating device, some method must be used for connecting the source voltage to it. This is done by using a pair of slip rings, with these corresponding to the commutator in a DC machine (see Figure 6-2). Thus,

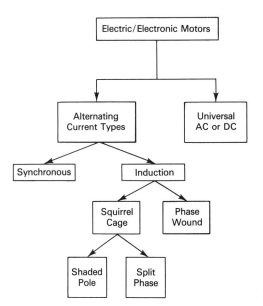

Figure 6-1 Relationships of AC motors.

TABLE 6-1. TYPES OF AC MOTORS

Universal
   Concentrated pole, non-compensated
   Distributed field, compensated
Repulsion
   Repulsion induction
   Repulsion start induction
Induction wound rotor
   Squirrel cage
   Single phase
   Split phase
   Polyphase
   Delta-connected polyphase
   Wye-connected polyphase
   Series
Hall effect
Brushless, coreless, slotless
Synchronous
Capacitor start
Shaded pole

the AC motor, like its DC counterpart, requires brushes, either carbon or metallized types.

Unlike the commutator, slip rings are continuous bands of metal. The slip rings, made of hard-drawn copper, are insulated from each other and from the shaft on which they are mounted.

The Universal Motor

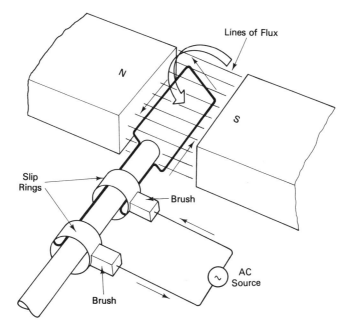

Figure 6-2 Basic AC motor.

THE UNIVERSAL MOTOR

The universal motor (Figure 6-3), sometimes known as a *concentrated-pole, non-compensated universal,* is so called since its source voltage can be either single-phase AC or DC. To make it suitable for DC input it uses a commutator instead of slip rings. The field-pole laminations, non-compensated type is not too costly to manufacture, since the field-pole laminations, the pole pieces, and the yoke are manufactured by a single machine punch.

The characteristics of this motor are listed in Table 6-2.

The characteristics of the universal motor are similar to those of the series-wound DC. The universal is also series wound, and has a high starting torque. Its speed depends on loading conditions, and in the absence of a load the armature will race. These motors are in the fractional horsepower range, extending from about 1/200 to 1/3 hp, though some have a rating of as much as 1/2 hp.

The armature is a wound type using a laminated core and having either straight or skewed slots. The motor can operate on AC or DC, but the motor speed is somewhat lower for AC under identical loading conditions.

Universal motors are often used in home appliances, including vacuum cleaners, food mixers, and tools. These operate at speeds ranging from 3,000 to 10,000 rpm.

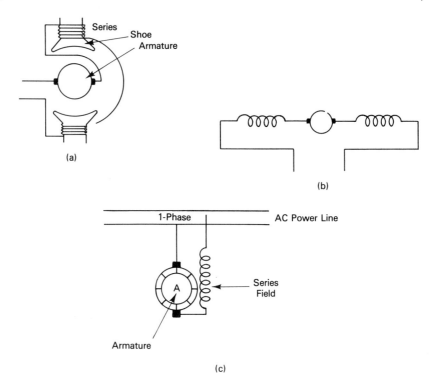

Figure 6-3 Connections for universal motor (a) and symbol (b); connection to single-phase AC power line (c).

TABLE 6-2. CHARACTERISTICS OF THE UNIVERSAL MOTOR

1. Speed is approximately the same whether using DC or AC.
2. Available in fractional hp sizes.
3. Field and armature coils are in series.
4. High starting torque.
5. Variable speed. Runs at high speed without load.
6. Direction of rotation can be reversed.
7. Brush holder is usually a fixed-position, non-adjustable type.
8. Universal motors are most often fractional hp types, in the range of 1/200 to 1/3.
9. Electrical power requirements from 4 to 250 W.
10. Typical rotation speed is 3,500 rpm.
11. Motor is equipped with two field windings.
12. Motor may have built-in fan mounted on the shaft.
13. Sometimes equipped with reduction gear to supply higher torque at lower speeds.
14. Speed can be varied by adjusting frequency of input voltage.
15. Radio-frequency interference (RFI) can be caused by brush sparking.

The Universal Motor

Speed Control

The speed can be controlled by putting a variable resistor in series with the "hot" side of the AC power line, as indicated in Figure 6-4. Another speed control technique is to use a tapped inductor as in Figure 6-5. This arrangement has the advantage of a lower operating cost, since there is less of a power loss using the tapped coil, but the initial cost may be higher than that of the variable resistor.

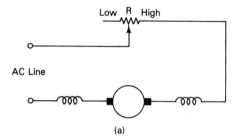

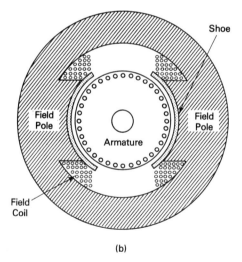

Figure 6-4 Speed control for a universal motor: circuitry (a), motor structure (b). The rheostat, R, is simple and effective, but there is a power loss in this component.

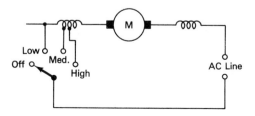

Figure 6-5 Universal motor with tapped field winding.

Compensated Series Wound

This is a variation of the straight-series type universal motor and has its field coils positioned in slots. This results in better speed regulation and a higher starting torque. The compensated series wound is sometimes used in motors having a rating of more than 1/3 hp.

Distributed Field Compensated

This universal motor type has its field coil windings mounted in slots cut in an iron core. This results in better speed regulation and a higher starting torque. It has a better performance capability than the non-compensated type.

PLUG AND SOCKET SYMBOLS

Some motors are connected directly into a fuse box. Connections and disconnections of the power cord can be made to a terminal board mounted on the motor frame. Lower horsepower motors, such as those used in the home, have their power lines wired to a male plug. Symbols for these plugs are illustrated in Figure 6-6(a). Some plugs are grounded types; others are not. There is no standardization for motor plug symbols. Symbols for sockets are shown in Figure 6-6(b).

SINGLE-PHASE AND POLYPHASE MOTORS

Single-phase AC motors include repulsion, induction, series, capacitor, and synchronous motors. There are also some variations under these general headings. A polyphase motor is one whose AC source voltage consists of more than one phase. This polyphase voltage can be supplied by the power line or else by single-phase

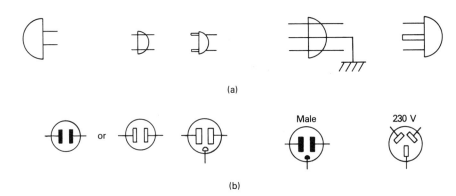

Figure 6-6 Motor plugs (a); sockets (b).

input that can subsequently be modified to polyphase. If a motor is connected to a single-phase voltage source, it doesn't necessarily mean the motor is a single-phase type.

REPULSION MOTORS

There are various kinds of repulsion motors, including repulsion, repulsion induction, repulsion start induction, and the induction wound rotor.

Repulsion Motor Characteristics

The characteristics of repulsion motors are listed in Table 6-3.

Repulsion-Start Induction

This motor has a wound rotor, a commutator, and brushes. The purpose of the commutator/brush combination is to work as an aid in starting the rotor. When it reaches about 75% of its running speed, a centrifugal governor lifts the brushes from the commutator and puts a short-circuiting ring around it. The motor then runs as though equipped with a squirrel-cage rotor. Table 6-4 lists the operating characteristics.

TABLE 6-3. CHARACTERISTICS OF REPULSION MOTORS

1. Rotation reversal can be obtained by shifting the brushes left or right of neutral.
2. The brushes are connected so as to short-circuit selected pairs of armature coils, with the short circuiting a technique used in squirrel cage motors.
3. The field coils are wired in series and are connected to a single phase AC line.
4. The armature obtains its current flow by induction, a result of the magnetic flux around the field coils. The motor functions like a step-down power transformer having a shorted secondary winding.
5. It is characterized by having a high starting torque, a moderate starting current, and a low power factor, but not at high speeds.

TABLE 6-4. CHARACTERISTICS OF THE REPULSION-START INDUCTION MOTOR

1. Starting torque is high and can be as much as four times the running torque.
2. Starting current is about three times the running current.
3. Horsepower output ranges from 1/6 to 20.
4. Source voltage can be 120 to 240. The motor must be designed for a specific input.
5. Speed availability is 1,200, 1,800, and 3,600 rpm.
6. Motor is reversible. Achieved by setting of brushes.

Repulsion Induction

This motor has a double armature consisting of a wirewound and a squirrel cage. The squirrel cage is positioned below the wound armature, which is put into slots. The motor uses a commutator. The wound armature is often a lap-wound type and is cross-connected. The characteristics of this motor are listed in Table 6-5.

Table 6-6 lists the fuse ratings.

Figure 6-7 shows the structure of the armature, motor details, and the symbol. The inner part consists of the squirrel cage. Above this are slotted laminations for an armature winding. While the slots shown in this drawing are straight, a skew arrangement is much more common. The wound armature may be a cross-connected lap-wound type. The advantage of the skewed slot is that a greater length of the active element of the wound armature is in the slot, thus increasing the inductance of the winding. It also supplies the same starting torque independently of the armature position.

The repulsion-induction motor has a number of advantages. It has a dual-voltage input capability, with operating characteristics comparable to those of the compound-wound DC motor. The motor has a high starting torque, and its operating speed remains fairly constant. The direction of armature rotation can be reversed by shifting the brushes approximately 15°. The amount of brush shift and its direction may be indicated on either the brush holder or on the motor frame.

TABLE 6-5. CHARACTERISTICS OF THE REPULSION-INDUCTION MOTOR

1. Dual-voltage input capability.
2. Operating characteristics similar to a DC compound wound.
3. High starting torque.
4. Fairly constant operating speed.
5. Armature rotation can be reversed by brush shifting.
6. May have compensating windings for increased power factor.

TABLE 6-6. FUSE CURRENT RATINGS FOR REPULSION-INDUCTION MOTORS

Horsepower	Current rating (amperes)
1/8	4
1/6	4
1/4	6
1/3	8
1/2	10
3/4	15

The Squirrel-Cage Concept

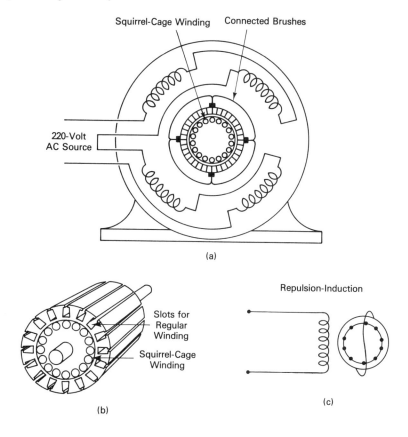

Figure 6-7 Repulsion-induction motor may have cross-connected lap-wound armature coils (a); structure of the armature (b); symbol (c).

In some motors there are limit stops to make the positioning of the brushes easier. This motor does not require a centrifugal switch.

Some repulsion induction motors are equipped with compensating windings for increasing their power factor (see Figure 6-8).

THE SQUIRREL-CAGE CONCEPT

The magnetic field surrounding the armature is based on the product of the number of armature coil turns and the current flowing through those turns, plus the permeability of the iron core. The number of turns and the amount of current flow are electrically and mechanically limited, the number of turns being limited by the wire size, which also limits the amount of current flow.

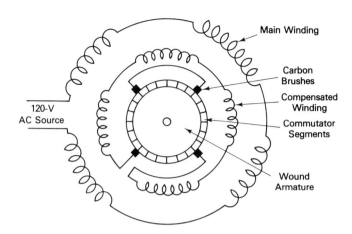

Figure 6-8 Use of compensating winding.

A squirrel-cage motor can be regarded as a step-down transformer (Figure 6-9). The smaller the number of secondary turns the greater the current flow induced in them by the primary winding. The maximum current flow is obtained when the secondary is shorted.

The primary of the transformer is equivalent to the field coils of a motor; the secondary comparable to the armature. Instead of a winding for the secondary, a squirrel cage can be used. There are various physical forms of the secondary, but it can consist of flat strips, bars, or rods short circuited at their ends by a copper or brass ring in large motors or aluminum in smaller ones.

The segments of the squirrel cage form a cylindrical cage, with the bars parallel to the motor shaft. The squirrel cage is illustrated in Figure 6-10. Because of the current induced in the squirrel cage, motors using it sometimes come under the heading of induction types. Table 6-7 lists the characteristics of the squirrel cage, sometimes called an *amortisseur winding*.

Some motors are equipped with a double squirrel cage. Ordinarily, the single squirrel cage has a very low resistance and high inductance. The second cage, positioned on top of the first cage but separated from it, has a higher resistance and a lower inductance. Figure 6-11 illustrates the symbol.

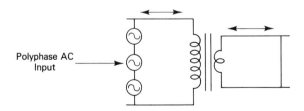

Figure 6-9 Squirrel-cage motor behaves like a transformer having a single-turn shorted secondary.

Single-Phase Motors

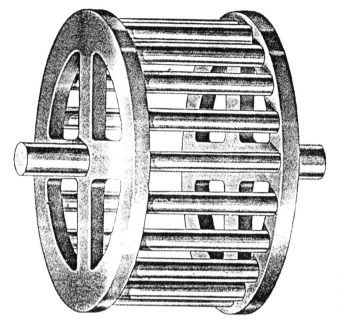

Figure 6-10 View of squirrel-cage rotor showing the end pieces, the copper connecting rods, and the shaft.

TABLE 6-7. CHARACTERISTICS OF SQUIRREL-CAGE MOTORS

1. Squirrel-cage motors do not use a commutator or brushes.
2. Electrical energy in the squirrel cage is obtained from the field coils by induction. There are no connections between the field coil and squirrel cage.
3. Polarity of the squirrel cage is always opposite that of the field coils.
4. Speed is determined by number of pairs of poles, frequency of AC source and to some extent, the load.
5. A squirrel cage imposes a heavy load on the field coils so these coils must be able to carry a strong current.
6. Squirrel cage motors require a polyphase source voltage.
7. High starting torque.
8. Good efficiency.

SINGLE-PHASE MOTORS

Motors can be characterized by the type of source voltage, such as DC or AC, and in the case of AC, by whether that input is single phase or polyphase. Single-phase motors can include repulsion, induction, series, capacitor, and synchronous types. There are also some variations under these general headings. Single-phase motors are mostly those used for in-home use; polyphase for industry, although there are some exceptions. Single-phase power can be modified so that it assumes the characteristics of polyphase power. Table 6-8 lists the power line lengths for single-phase motors ranging from 1/4 to 10 hp.

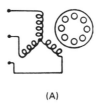

Figure 6-11 Symbol for a squirrel-cage motor. This is for a three-phase stator winding.

POWER-LINE LOSSES

The entire current of a motor flows through the power lines connecting the motor to its power source. The IR drop along the lines and the I^2R losses can affect motor efficiency. Furthermore, while the line source voltage is often assumed to be 120 V, this is its RMS value and is based on the peak voltage. That peak can fluctuate depending on external loads, thus RMS can often be well below 120 V. In practice the RMS can be 110 V, or even lower.

The chart in Table 6-9 assumes an input of 115 V RMS. Columns A, B, and C represent conditions for 1%, 2%, and 3% voltage drops. Divide these by the number of amperes flowing in the motor circuit to obtain the distance in feet to the power source for the percentage of voltage drops indicated.

Power-Line Frequency

Unlike line voltage, power-line frequency is remarkably constant. While a 60-Hz line can have a frequency shift of plus or minus a half-cycle, over a 24-hr period the frequency averages 60 Hz.

A similar listing for 230-V motors appears in Table 6-10. The 230-V supply is in RMS and is subject to the same restrictions that apply to Table 6-9. The numbers supplied in columns A, B, and C are twice the corresponding numbers in Table 6-9. These numbers are to be divided by the number of amperes flowing in the motor circuit to determine the distance in feet to the power source.

In either case, the voltage drop can be lowered by reducing the wiring distance to the source, or by using a lower gauge number, or by a combination of both.

Table 6-11 approaches the problem of power line wire size in a different way. The data shown are for a one-way distance to the motor, that is, these indicate the maximum one-way distance in feet that each size of wire will carry current for motors with different horsepower ratings. These numbers represent a 3% voltage drop along the power line.

These data assume that the motor will not be operated in an overload condition. The larger the overload, the greater the current demand, resulting in a larger voltage drop.

Voltage drops and power losses can also occur at the connecting terminals for the power lines at the motor and also at the fuse box. If these points are hot, they will represent power losses and voltage drops.

TABLE 6-8. POWER-LINE LENGTHS FOR SINGLE-PHASE MOTORS

Horsepower of motor	Volts	Approximate starting current (amperes)	Approximate full-load current (amperes)		Length of wire (feet)							
				Feet	25	50	75	100	150	200	300	400
¼	115	20	5	AWG	14	14	14	12	10	10	8	6
⅓	115	20	5.5		14	14	14	12	10	8	6	6
½	115	22	7		14	14	12	12	10	8	6	6
¾	115	28	9.5		14	12	12	10	8	6	4	4
¼	230	10	2.5		14	14	14	14	14	14	12	12
⅓	230	10	3	AWG	14	14	14	14	14	14	12	10
½	230	11	3.5		14	14	14	14	14	12	12	10
¾	230	14	4.7		14	14	14	14	14	12	10	10
1	230	16	5.5		14	14	14	14	14	12	10	10
1½	230	22	7.6		14	14	14	14	12	10	8	8
2	230	30	10	AWG	14	14	14	12	10	10	8	6
3	230	42	14		14	12	12	12	10	8	6	6
5	230	69	23		10	10	10	8	8	6	4	4
7½	230	100	34		8	8	8	8	6	4	2	2
10	230	130	43		6	6	6	6	4	4	2	1

TABLE 6-9. POWER-LINE VOLTAGE DROPS FOR 115-V INPUT

Wire size	A 1% Voltage drop	B 2% Voltage drop	C 3% Voltage drop
14	214	428	642
12	340	680	1,020
10	540	1,080	1,620
8	858	1,716	2,574
6	1,342	2,684	4,026
4	2,124	4,248	6,372
3	2,683	5,366	8,049
2	3,395	6,790	10,185
1	4,264	8,528	12,792
0	5,392	10,784	16,176

TABLE 6-10. POWER-LINE VOLTAGE DROPS FOR 230-V INPUT

Wire size	A 1% Voltage drop	B 2% Voltage drop	C 3% Voltage drop
14	428	856	1,284
12	680	1,360	2,040
10	1,080	2,160	3,240
8	1,716	3,432	5,148
6	2,684	5,368	8,052
4	4,248	8,496	12,744
3	5,366	10,732	16,098
2	6,790	13,580	20,370
1	8,528	17,056	25,584
0	10,784	21,568	32,352

Current Requirements for Single-Phase Motors

The data in Table 6-12 indicate the approximate amount of operating current required by single-phase motors with source voltages of 115, 230, or 440 V. These source voltages are RMS values and can vary from the amounts stated. The current figures can also change accordingly and are also dependent on the amount of motor loading. Consequently, the motor current data should be regarded as guidelines rather than precise amounts.

The currents listed are full load, with the motors running at their designated speeds and with normal torque. To obtain the full-load currents of 200- and 208-V motors, increase the corresponding 230-V motor full-load currents by 10% and 15%, respectively.

TABLE 6-11. WIRE SIZE VERSUS DISTANCE TO POWER SOURCE

Horse-power	Volts	No. 14	No. 12	No. 10	No. 8	No. 6	No. 4	No. 2	No. 0
1/4		140	220	350	560	890	1,400	2,300	3,600
1/3		110	170	260	420	660	1,100	1,700	2,700
1/2	115	90	140	220	350	560	890	1,400	2,200
3/4		60	100	160	250	400	640	1,000	1,600
1			80	130	200	320	450	800	1,300
1/4		560	890	1,400	2,250	3,600	5,700	9,000	
1/3		420	660	1,050	1,670	2,600	4,200	6,700	
1/2		350	550	880	1,400	2,200	3,500	5,600	8,900
3/4		250	400	640	1,010	1,600	2,600	4,200	6,500
1		200	320	500	800	1,300	2,000	3,200	5,100
1 1/2	230	140	220	350	560	900	1,400	2,300	3,600
2		110	170	270	430	690	1,100	1,800	2,800
3				190	310	480	860	1,200	1,900
5					190	290	470	740	1,200
7 1/2						210	320	520	820

TABLE 6-12. CURRENT REQUIREMENTS OF SINGLE-PHASE MOTORS

Horsepower	115 V	230 V	440 V
1/6	4.4	2.2	—
1/4	5.8	2.9	—
1/3	7.2	3.6	—
1/2	9.8	4.9	—
3/4	13.8	6.9	—
1	16	8	—
1 1/2	20	10	—
2	24	12	—
3	34	17	—
5	56	28	—
7 1/2	80	40	21
10	100	50	26

Current versus Electrical Power Input

Another method for determining the current demand of a motor is shown in Table 6-13. This table is intended for motors having a 115-V, single-phase power input, and assumes the motor has a unity power factor or very close to it.

The horsepower ratings in the column at the left include fractional types. The row across the top, starting at 0 and ending with 9 indicates whole number amounts of power in kilowatts. The first column at the left, starting at 0 and ending with

TABLE 6-13. CURRENT REQUIREMENTS OF 120 V, SINGLE-PHASE MOTORS

Kilowatts	0	1	2	3	4	5	6	7	8	9
0	—	8.70	17.39	26.09	34.78	43.48	52.17	60.87	69.56	78.26
0.1	0.87	9.57	18.26	26.95	35.65	44.35	53.04	61.74	70.43	79.13
0.2	1.74	10.43	19.13	27.82	36.52	45.22	53.91	62.61	71.30	80.00
0.3	2.61	11.30	20.00	28.69	37.39	46.09	54.78	63.48	72.17	80.87
0.4	3.48	12.17	20.87	29.56	38.26	46.96	55.65	64.35	73.04	81.74
0.5	4.35	13.04	21.74	30.43	39.13	47.83	56.52	65.22	73.91	82.61
0.6	5.22	13.91	22.61	31.30	40.00	48.75	57.39	66.09	74.78	83.48
0.7	6.09	14.78	23.48	32.17	40.87	49.56	58.26	66.96	75.65	84.35
0.8	6.96	15.65	24.35	33.04	41.74	50.43	59.13	67.83	76.52	85.22
0.9	7.83	16.52	25.22	33.91	42.61	51.30	60.00	68.70	77.39	86.09

0.9 represents fractional amounts of kilowatts. Thus to find the current demand of a single phase motor having an input power of 7.3 kW, locate digit 7 in the top row and move downward. Then find 0.3 in the first column at the left and move toward the right. The meeting point will be at 63.48 A.

SHADED-POLE MOTOR

The shaded-pole motor is somewhat like a modified single-phase induction type, except for a change made in one or more of its field poles. Sometimes the motor is referred to as a *shaded-pole induction type*.

Figure 6-12 shows the basic construction of a four-pole shaded-pole motor. The rotor is a squirrel cage, and the field coils are series wound and connected to 120-V single-phase AC. There is no commutator, no slip rings, and no brushes.

The shading coil, a single bare-wire shorted turn or an uninsulated copper ring, is placed around one tip of each field pole, as indicated in the drawing. The effect is to produce a very weak starting torque, thus motors of this type are intended only for small loads. The starting torque is in the range of 50% to 80% of the full speed torque at the moment the motor starts. The torque then rises momentarily to 120% at 90% of full speed, dropping to normal torque at operational speed.

Shaded-pole motors have low efficiency and poor speed regulation, and are not generally considered as reversible types, although reversing is possible. These motors are fractional-horsepower types with 1/25 hp common.

Figure 6-12 shows the arrangement of a four-pole shaded-pole motor. The shading coils, copper rings in this example, have no connection to the power line and receive their excitation by induction only. Figure 6-13 illustrates the symbol. It shows two coils, with the larger representing the field coils, the smaller the shaded poles. The circle in the center, enclosing a number of much smaller circles, is the squirrel-cage rotor. The lower symbol is easier to draw.

Shaded-Pole Motor

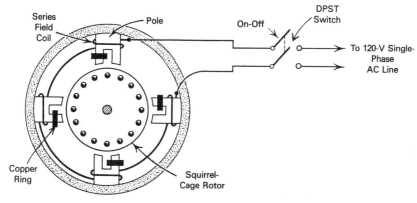

Figure 6-12 Four-pole shaded pole motor.

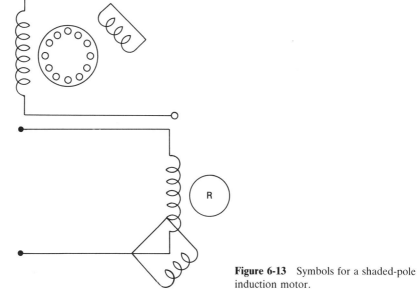

Figure 6-13 Symbols for a shaded-pole induction motor.

Summary of the Characteristics of the Shaded-Pole Induction Motor

1. Load: Designed for easy starting
2. Starting Torque: Low. Can be less than or equal to running torque
3. Horsepower: Ranges from 1/20 to 1/4 hp
4. Phase: Single
5. Source Voltage: 120 V
6. Starting Current: Low
7. Speed: 900, 1200, 1800, and 3600 rpm, constant
8. Reversibility: No

9. Noise and Vibration: Medium
10. Efficiency: Low to Medium
11. Uses: Motor-operated toys, hair dryers, small fans, electric clocks, advertising displays
12. Speed Control: Can be obtained by series resistance or iron-core inductor.
13. Commutator: None
14. Brushes: None
15. Slip Rings: None
16. Field Coils: Series wound

SYNCHRONOUS MOTORS

A synchronous motor is one in which the armature rotates in step with the rotating magnetic field of its field coils. It is from this behavior that the motor obtains its name, since its armature is synchronized to some multiple of the power-line frequency. It maintains its speed except when it is heavily overloaded; thus it is regarded as a constant-speed motor.

The speed of rotation of the armature in revolutions per minute can be determined by the number of pairs of poles and the line frequency:

$$rpm = \frac{f \times 60}{p}$$

where

rpm = revolutions of the armature per minute,

f = the frequency of the sine wave AC input voltage, and

p = the number of poles, always an even number such as 2, 4, 6, 8, or more and is a constant

Thus, for a four-pole motor operating from the 60-Hz power line (with Hz representing cycles per second),

$$rpm = \frac{60 \times 60}{4} = \frac{3,600}{4} = 900 \; rpm$$

See Table 6-14.

Self-Starting Induction-Reaction Synchronous Motor

The self-starting induction-reaction synchronous motor is still another type and is a combination of induction and synchronous motors. It features a shaded-pole field that is slotted, with six slots typical. It is these salient poles that supply the reaction effect. The armature isn't a wound type but is a squirrel-cage arrangement.

Split-Phase Motors

TABLE 6-14. SYNCHRONOUS SPEEDS IN RPM

Number of poles	60 Hz	50 Hz	40 Hz
2	3,600	3,000	2,400
4	1,800	1,500	1,200
6	1,200	1,000	800
8	900	750	600
10	720	600	480
12	600	500	400
14	514.2	428.6	343
16	450	375	300
18	400	333.3	266.6
20	360	300	240
22	327.2	272.7	218.1
24	300	250	200
26	277	230.8	184.5
28	257.1	214.2	171.5
30	240	200	160
32	225	187.5	150
34	212	176.5	141.1
36	200	166.6	133.3
38	189.5	157.9	126.3
40	180	150	120
42	171.5	142.8	114.2
44	163.5	136.3	109
46	156.6	130.5	104.3
48	150	125	100
50	144	120	96
52	138.5	115.4	92.3
54	133.3	111.1	88.9
56	128.6	107.2	85.7
58	124.1	103.5	82.8
60	120	100	80
62	116.1	96.8	77.4
64	112.5	93.7	75
66	109	90.8	72.7

SPLIT-PHASE MOTORS

A single-phase motor can be made into a polyphase type through the use of inductance or capacitance. With an inductor the current lags the applied voltage; with a capacitor it leads. Ideally, the phase separation is 90° but in actual practice it is less than this.

Split-phase motors operate from a single-phase power line, with the phase shifting technique supplied by the motor. Split-phase motors are used where only single-phase power is available. One of the requirements of split-phase operation is that the motor must be self-starting.

Self-Starting Techniques for Split Phase

Figure 6-14(a) is the wiring diagram for one type of split-phase motor. The rotor is a squirrel cage and has no external connections, developing its operating voltage by induction. The stator field coils are a two-part device consisting of the main field coils and a pair of start windings. The main windings are made of relatively low-gauge copper and are positioned at the bottom of their slots, thus increasing their overall inductance. The main windings, sometimes called the *running windings* or the *concentrated windings*, are in series and are connected directly across the single-phase AC power line.

The starting or auxiliary windings are made of finer wire (higher-gauge wire) and are positioned near the top of their slots. Use of higher-gauge wire means a higher resistance; the position in the slot means a lower value of inductance. The starting windings are designed to be mainly resistive, the main windings inductive. As a result the current flow through these windings is out of phase, not by 90° but reasonably close to that number. Physically, the main stator windings and the starting windings are at a 90° angle.

The starting windings are also in series with each other, with this series combination connected to the AC power line.

The result of this design is a weak rotating magnetic field, which provides a low starting torque. As the armature continues to turn, it will reach about 80% of its synchronous speed. At that time the centrifugal switch will open, and the motor will continue to operate as a single-phase induction type. The centrifugal switch is attached to the rotor shaft and uses weights that are thrown outward by centrifugal force. Figure 6.14(b) shows the symbol for this motor.

The starting winding is not intended to carry a full current load continuously, and so the centrifugal switch is used to prevent starting coil burnout. If the starting coil is open, the armature can be given a swift spin manually, and the armature may continue to turn.

Reversing Direction

This split-phase motor will run in the started direction. To reverse operating direction, either the starting or running coil connections to the power line must be transposed (but not both).

Starting versus Running Current

The low starting speed of the split-phase motor results in a large field and starting coil current (larger through the field winding), and this can be more than 15 A. At full speed this drops to about 4 to 5 A. This action requires fusing by using

Split-Phase Motors

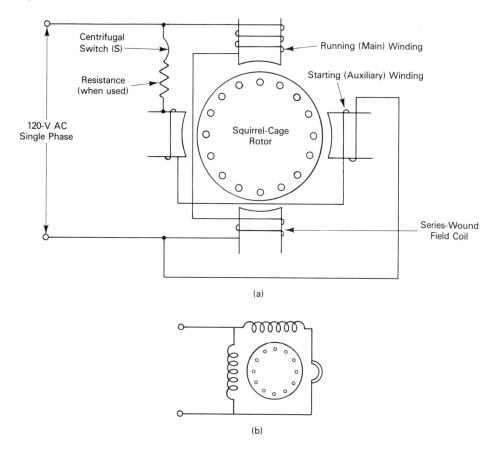

Figure 6-14 Split-phase induction motor (a); symbol (b).

time-lag fuse types or circuit breakers. These fuses can also blow if the motor is overloaded.

Armature RPM

The speed of this motor's rotating magnetic field, called the *synchronous speed*, is slightly higher than its actual speed. The difference between the two is the slip. The synchronous speed for a four-pole motor operating from 60 Hz is 1800 rpm for a four-pole motor. If the slip is 75 rpm, the actual speed will be

$$\text{actual speed} = \text{synchronous speed} - \text{slip}$$

Slip can vary slightly caused by increasing the motor load or by a decrease in the line voltage.

Speed Variation

The inductive split phase motor can perform at two or three different speeds provided multiple windings are used. Figure 6-15(a) shows the circuitry; Figure 6-15(b) shows the symbol.

Capacitor Split-Phase Motors

A more desirable type of split-phase motor uses a capacitor to obtain the required phase shift. Figure 6-16(a) shows the wiring diagram.

This motor has a pair of windings that are physically separated by 90° with the windings connected to a single-phase, 60-Hz power line. One of the windings is in series with a capacitor.

This motor may or may not be equipped with a centrifugal switch, depending on load requirements. The size of the capacitor in terms of capacitance can be large enough to require the use of electrolytic types with the capacitor mounted on the motor frame. Some of these motors are equipped with a pair of capacitors, with these connected in parallel so as to supply a greater capacitance. However, the total capacitance may be too large during normal load and running conditions, so the unit may be equipped with a centrifugal switch to open the connections to

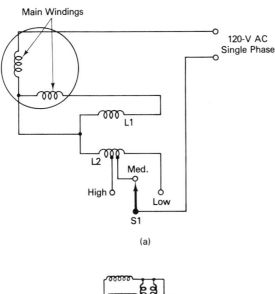

Figure 6-15 Multi-speed split-phase inductive motor (a); symbol (b).

Split-Phase Motors

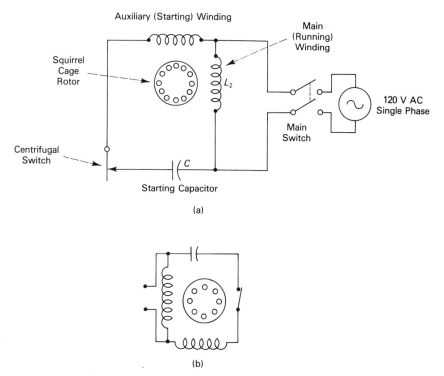

Figure 6-16 Capacitor split phase motor (a); symbol (b).

one of the capacitors when the motor approaches its full speed. The two capacitors do not have the same capacitance. The one with the larger amount is the one removed by the centrifugal switch.

Capacitor split-phase motors are used for home appliances such as washing machines, commonly in the 1/4-hp size. The motor can be used only on AC. A DC input would probably burn out the windings.

Summary of the characteristics of the capacitor split-phase motor.

1. Load: Difficult starting load
2. Starting Torque: 300% to 400% of the running torque
3. Horsepower: 1/6 to 10 hp
4. Phase: Single
5. Source Voltage: 120 to 240 AC
6. Speed: 900, 1200, 1800, 3600 rpm
7. Reversibility: Yes
8. Uses: Machine tools, grinders, blowers, ventilating fans

Symbol. Figure 6-16(b) illustrates the symbol for the capacitor split-phase motor. Most of the symbol is fairly standard, such as the squirrel-cage rotor and the running coil. However, there are several variations for the capacitor and the centrifugal switch.

Capacitor-Start Two-Speed Motor

Figure 6-17(a) is the circuit of a two-speed motor, a capacitor-start type. The switch shown at the top left has two positions: low speed and high speed. The switch at the top right is a centrifugal unit, with the letter *S* to indicate starting torque, the letter *R* an abbreviation for running torque. There are two running windings, 1 and 2, with running winding 1 used for low speed and running winding 2 for high speed. The capacitor is in the circuit when the motor starts but is disconnected when the motor approaches its running speed. This could be the type of motor used for a fan when two speeds are wanted for a specific, unvarying load.

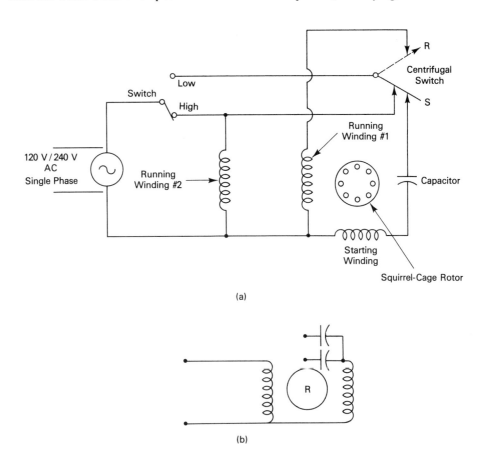

Figure 6-17 Two-speed capacitor start motor (a); symbol (b).

Split-Phase Motors

Two-Capacitor Motor

The two capacitor motor, as its name implies, uses a pair of capacitors, with the value of the capacitors selected for be   tarting and running torque. As shown in Figure 6-18 the two capacitors are in shunt with one of them in series with a centrifugal switch. When capacitors are wired in parallel, the total capacitance is the sum of the individual capacitances. This is the condition that prevails when the motor is first turned on. After the armature gets close to its final running speed, the switch opens and only the running capacitor remains.

The two capacitors are different types. The running capacitor—the one that remains permanently connected—is a sealed paper type; the starting capacitor is an AC electrolytic.

Permanent Capacitor Motor

For some applications it isn't necessary for the motor to have a centrifugal switch; thus the start capacitor and the auxiliary winding remain permanently connected. This can be done for motor operations in which the load remains constant. The torque is not as good as in those motors using a centrifugal switch.

The capacitor is wired in series with the auxiliary winding and so behaves like a series-tuned L-C circuit. The inductive reactance of the coil and the capacitive

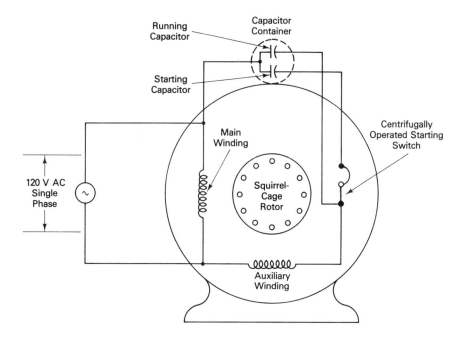

Figure 6-18 Two-capacitor motor.

reactance of the capacitor are 180° out of phase and have a canceling effect. If the reactances are exactly equal, they will cancel completely, and so the net reactance of the combination will be due entirely to the resistance in that circuit. As such, the resistance will be 90° out of phase with the reactance of the main winding. However, this phase difference will be this amount only at a specific AC input frequency to the motor. The circuit is equipped with a centrifugal switch, and when the armature is close to synchronous speed, will open, removing the capacitor.

The capacitor, usually known as a *starting capacitor*, is often a dry AC electrolytic. Because the operating frequency of the motor is fairly low (60 Hz), large values of capacitance are required. This can be most conveniently supplied by this type of capacitor. The capacitance tolerance of such capacitors is fairly large, so obtaining a full 90° phase shift is seldom practical. Fortunately it isn't a necessity, and in practice lower values of phase shift are used.

One of the chief advantages of the capacitor-start motor is its high starting torque. In this respect it is much higher than that of the split-phase motor. Consequently this motor is desirable when the load is such that a higher starting torque is necessary. Furthermore, the starting current is low and motor operation is quiet. The motor is also noted for its not interfering with radio or television reception, making it favored as a motor for use in refrigerators.

Autotransformer Effect

Unlike other power transformers, which have individual primary and secondary windings not connected physically, an autotransformer (Figure 6-19) consists of a single tapped winding. A voltage stepup is obtained because of the turns ratio: the secondary portion has more turns than the primary.

An autotransformer can be used in conjunction with the starting capacitor of the capacitor motor (Figure 6-20). The autotransformer does not change the AC line input frequency, but does step up the voltage applied to the capacitor, above that supplied by the line. This produces an effect equivalent to increasing the capacitance. As a result, this makes possible the use of a capacitor having a smaller value of capacitance. The capacitor, however, must have a working voltage rating capable of withstanding the higher value of voltage. With this setup, the unit still

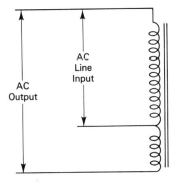

Figure 6-19 Autotransformer. The unit can have a multi-tap output.

Split-Phase Motors

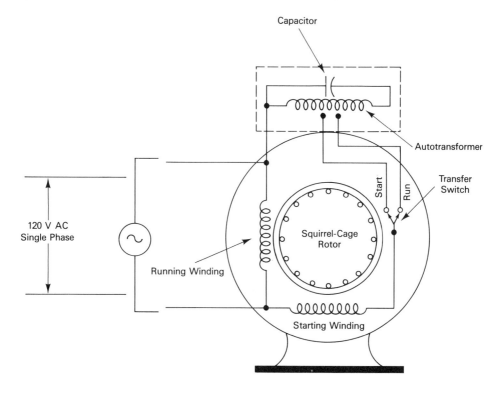

Figure 6-20 Motor using starting capacitor and autotransformer.

works as a capacitor motor. When the centrifugal switch cuts out, the motor runs as an induction type.

Split-Capacitor Motor

This version of capacitor motor is diagrammed in Figure 6-21. The capacitor consists of two units, one of which is a starting type, the other a running unit, with both housed externally on the motor frame in a metal container. When the motor reaches its running speed, the starting capacitor is disconnected by the centrifugal switch.

The motor also has a pair of windings, one of which is the main, the other the start winding. The start winding is also disconnected from the circuit when the centrifugal switch opens.

Summary of the characteristics of the split-capacitor motor.

1. Load: Designed for easy starting.
2. Starting Torque: Low. It can be 50% to 100% of the running torque.
3. Starting Current: Low.
4. Horsepower: 1/20 to 1 hp.

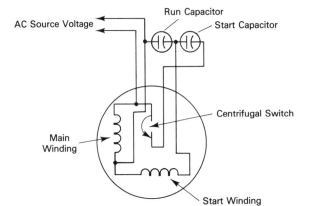

Figure 6-21 Split-capacitor motor.

Summary of the Characteristics of Split-Phase Motors

1. Split-phase motors are fractional-horsepower types.
2. Mechanical power output ranges from 1/50 to 1/3 hp.
3. Low starting torque and is a fraction of normal full-speed torque.
4. Good speed regulation.
5. Motor is suitable for easy-to-start light loads.
6. Requires large starting current, which can be six to eight times the full-speed running current.
7. Motor is equipped with two starting windings.
8. Uses squirrel-cage rotor.

CENTRIFUGAL SWITCHES

Motors, such as split-phase inductor and capacitor types, can use a switch to open the connectors to the phase-shifting inductor or capacitor, often when the motor is close to about 75% of its final running speed.

There are various centrifugal switch designs, one of which is shown in Figure 6-22(a). It consists of a pair of ball weights held in position by a spring. This holds the contacts together, supplying a path for the starting coil. When the armature reaches close to its final running speed, the centrifugal force exerted on the ball weights moves them outward, separating the contacts and interrupting the starting coil current.

Another type of centrifugal switch is in Figure 6-22(b). The drawing at the left shows the condition at the moment the motor switch is turned on. The contacts are in their "on," or closed, position. As the motor approaches operating speed, centrifugal force drives the flyweights outward, as indicated in the drawing at the right. When the flyweights overcome the spring tension, the contacts open, dis-

Centrifugal Switches

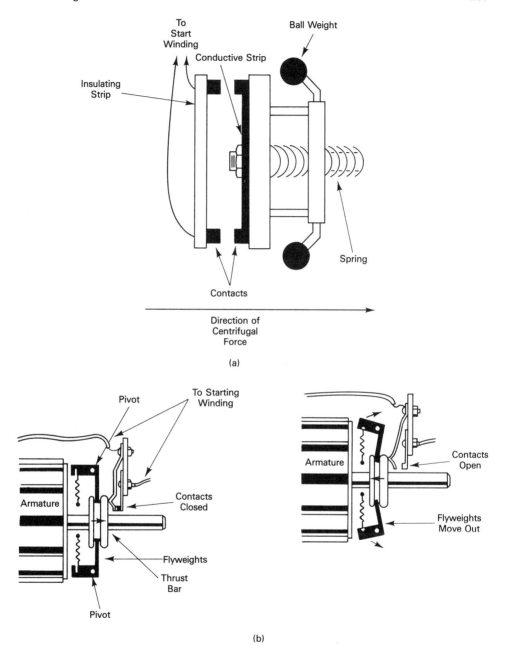

Figure 6-22 Centrifugal switches. Ball-weight type (a); flyweight type (b).

connecting the circuit of the start winding, if a coil is used, or the connections to the capacitor of a capacitor-start unit.

POLYPHASE MOTORS

A polyphase motor is one whose source input must be a polyphase voltage. While there are two-phase motors, polyphase types ordinarily require a three-phase voltage input.

A polyphase voltage system can be regarded as a combination of two or more single-phase systems. While the separation of each phase in polyphase is measured in degrees, each voltage wave is time-separated, that is, the first voltage wave starts before the second, the second before the third. Following the third, the action repeats.

Advantages of Polyphase

Single-phase AC is actually a pulsating voltage, and twice during each cycle no power is delivered, since the voltage and current are zero at those times. Polyphase is also pulsating, but since there are more pulses per unit of time, delivery of power is smoother.

The construction of polyphase motors is simpler than for single phase. Polyphase motors are noted for having a higher efficiency and for needing fewer repairs. Single-phase motors are generally used for fractional-horsepower types; polyphase for larger motors. The power rating of a polyphase motor is higher than that of single phase. However, the fact that an existing voltage source is polyphase does not preclude the use of single-phase motors, since these can be operated from any one of the phases. The converse isn't true unless some modification is made to the power source.

Polyphase power can be used on practically any constant-speed polyphase motor. It is intended mostly for commercial applications, since the in-home power source is generally 120 V, single phase. Most polyphase motors are usually made for 220 V, but are available for other source voltages, such as 440. Some motors are dual-voltage types and can accommodate an input of either 220 or 440 V. Polyphase motors can also be multi-speed types.

Polyphase Motor Windings

The windings of a polyphase motor can be arranged in either a delta or wye, as indicated in Figure 6-23. These drawings are shown in this way to emphasize the phase relationships of the windings, not their physical arrangement. The wye is sometimes called a *star*. Figure 6-24 shows the relationship between wye and delta windings. These windings can be in an equivalent form, as indicated in the drawing.

Three-phase voltages can be represented as in Figure 6-25. While the horizontal line is usually indicated in degrees, it can also be numbered to represent

Polyphase Motors

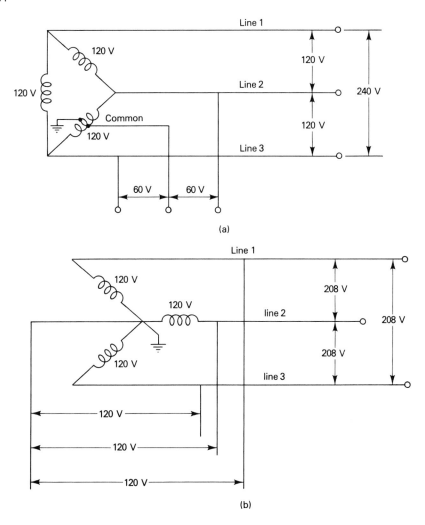

Figure 6-23 Delta connection (a); wye connection (b).

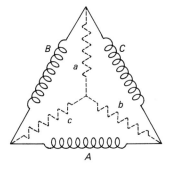

Figure 6-24 Equivalent relationships of delta and wye windings.

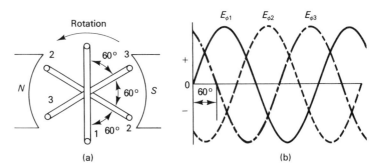

Figure 6-25 Independent generator armature coils, 1-1, 2-2, and 3-3, are spaced 60° apart (a) and produce three independent AC voltages, also spaced 60° apart (b).

units of time. Each phase has the same positive and negative amplitudes. At no time do all three voltages cross the zero line simultaneously, so power is always supplied and utilized. The magnetic fields produced by the polyphase currents rotate with each of these fields growing and decreasing in sequence.

Polyphase Induction Motors

Induction motors are also polyphase types and can be subdivided into two major groups: the wound rotor and the squirrel-cage rotor.

Wound-rotor induction motor. The rotor of this motor consists of groups of coils, which are designed so they have the same number of poles as the stator windings. While such motors are often polyphase types, they can be designed to work from a single-phase AC line. In all instances the purpose is to produce a rotating magnetic field.

Whether the rotor consists of a squirrel cage or is a wound-type winding, a strong current is produced by induction. The magnetic field accompanying this induced current reacts with the rotating magnetic field of the stator, thus causing the rotor to turn. This turning motion is slightly slower than that of the rotating magnetic field. To achieve the maximum induction effect, the air gap between the stator and the armature is made to have minimum clearance.

The three-phase induction motor has several advantages over the single-phase type. Since it has three equally spaced windings 120° apart the result is the induction of magnetic fields around the rotor that are out of phase with each other, producing the equivalent of a rotating magnetic field. The reaction of these fields with those of the stator windings produces armature rotation. No starting winding is needed, and the centrifugally operated switch isn't required. This motor is self starting.

Synchronous Speed

The speed of the rotating magnetic field of the stator winding is referred to as its *synchronous speed*.

Rotor Slip

The difference between the synchronous speed and the running speed of the rotor is known as *rotor slip*, or sometimes more simply as *slip*. The rotor speed of an induction motor is expressed by

$$N_2 = N_1(1 - s)$$

where

N_2 = the rotor speed,

N_1 = the synchronous speed (speed of the rotating field), and

s = the slip expressed as a decimal.

The slip of the rotor in an induction motor is

$$s = \frac{N_1 - N_2}{N_1}$$

where s is the slip speed, and is supplied as a decimal.

The rotational rate of the magnetic field is proportional to the frequency of the AC input and is inversely proportional to the number of poles. In terms of a formula it can be expressed as

$$N_1 = \frac{120f}{P}$$

where

P = the number of poles,

f = the frequency in hertz, and

N_1 = the speed of the rotating field.

The number of poles is their number per phase. A two-phase motor will have four poles, and a three-phase motor will have six.

The formula shows that the motor is independent of its load and is directly dependent on the frequency of the source voltage. Consequently, this type of motor can be regarded as a constant-speed device, assuming, of course, that the load does not exceed the breakdown torque of the armature.

The answers to problems supplied by these formulas is in decimal form. It can be changed to a percentage by multiplying the result by 100. In some instances the number 100 is included in the formulas so the result is directly in percentage.

Example:

An induction motor operating from a three-phase, 60-Hz power line operates at a speed of 1,730 rpm. What is the percentage slip of this motor?

Solution: The speed of the rotating magnetic field can be calculated by multiplying the line frequency by 120 and dividing the result by the number of poles. Thus

$$\text{Speed} = \frac{60 \times 120}{4} = 1{,}800 \text{ rpm}$$

The slip is equal to

$$\text{Slip} = \frac{1{,}800 - 1{,}730}{1{,}800} = 0.038$$

$$0.038 \times 100 = 3.8\%$$

There are two speeds involved here: the speed of the rotating magnetic field and the actual speed of the armature. The fact that there is a difference between these two does not mean the motor does not operate at a constant speed. In this example the motor speed is 1,730 rpm, and it remains at this value for its load range. However, there can be a modest difference between its full-load and no-load speeds.

Example:

A four-pole induction motor operating two-phase from a 60-Hz power line has a no-load speed of 1,790 rpm and a full-load speed of 1,750 rpm. What is the synchronous speed of this motor, the percentage of rotor slip under no-load conditions, and the amount of slip with the motor working at full load?

Solution: The synchronous speed is

$$N_1 = \frac{120 \times f}{P} = \frac{120 \times 60}{4} = 1{,}800 \text{ rpm}$$

The slip at no load is

$$\frac{1{,}800 - 1{,}790}{1{,}800} = 0.005 \text{ or } 0.5\%$$

The slip at full load is

$$\frac{1{,}800 - 1{,}750}{1{,}800} = 0.0277 = 2.77\%$$

Current Requirements of Polyphase Motors

Table 6-15 is a tabulation of the amount of current used by a 208-V, three-phase motor having a unity power factor. The first row across the top of this chart is in whole numbers of power input to the motor in kilowatts, starting with 0 and ending with 9. This row is used in conjunction with the first column, also headed in kilowatts. The amounts in this column are in decimal fractions, starting with 0 and ending with 0.9.

To determine the current taken by a motor whose input power is 4.5 kW, locate the number 4 in the top row. Move downward intercepting a line drawn horizontally from 0.5. The amount of current listed is 12.51 A.

Polyphase Motors

TABLE 6-15. CURRENT FOR 208-V, THREE-PHASE, UNITY-POWER-FACTOR MOTOR

Kilowatts	0	1	2	3	4	5	6	7	8	9
0		2.78	5.56	8.34	11.12	13.90	16.67	19.45	22.23	25.01
0.1	.28	3.06	5.84	8.61	11.39	14.17	16.95	19.73	22.51	25.29
0.2	.56	3.33	6.11	8.89	11.67	14.45	17.23	20.01	22.79	25.57
0.3	.83	3.61	6.39	9.17	11.95	14.73	17.51	20.29	23.07	25.84
0.4	1.11	3.89	6.67	9.45	12.23	15.01	17.79	20.56	23.34	26.12
0.5	1.39	4.17	6.95	9.73	12.51	15.28	18.06	20.84	23.62	26.40
0.6	1.67	4.45	7.23	10.00	12.78	15.56	18.34	21.12	23.90	26.68
0.7	1.95	4.72	7.50	10.28	13.06	15.84	18.62	21.40	24.18	26.96
0.8	2.22	5.00	7.78	10.56	13.34	16.12	18.90	21.68	24.46	27.23
0.9	2.50	5.28	8.06	10.84	13.62	16.40	19.18	21.95	24.73	27.51

This table can also be used for motors whose power factor is not unity. Divide the current that is listed by the power factor expressed as a decimal. Assume, for example, a motor whose input power is 3.8 kW. The table shows this as 10.56 A. For a motor having a 90% power factor, the current would be 10.56/0.9 = 11.73 A.

The current requirements of polyphase motors can not only be considered in terms of power input in kilowatts, but also with regard to output in terms of horsepower. Two charts are shown in Table 6-16. The first is for a 208-V, three-phase motor having an efficiency of 90% and an 85% power factor. The second table has the same specifications but is for a motor having an input source of 220 V.

The top row in both tables is horsepower in steps of 10. The first column is horsepower in unit steps. To find the current usage by a 220-V motor rated at 15 hp, locate the number 10 in the topmost row. Move downward to intersect a line drawn horizontally from the number 5. The current is 38.34 A. The other table can be handled in the same way for a 208 V motor.

There are so many variables involved in the determination of motor current that precise figures, supplied by tables, aren't possible. The best that can be expected is a reasonable approximation. The data supplied in Table 6-17 are for two- and three-phase squirrel-cage induction motors and for two- and three-phase slip-ring induction motors. The data has been compiled from average values of these motors. Variations of 10% above or below these figures are possible. The power factor of a three-phase motor can be calculated from

$$pf = \frac{P}{\sqrt{3}\,(E_L)(I_L)}$$

where

pf = the power factor,

P = the power delivered to the motor,

TABLE 6-16. HORSEPOWER-TO-AMPERES CONVERSION

208 V, Three-phase, 90% efficiency, 85% power factor

Horsepower	0	10	20	30	40	50	60	70	80	90
0		26.96	53.92	80.89	107.8	134.8	161.8	188.7	215.7	242.7
1	2.69	29.66	56.62	83.58	110.5	137.5	164.5	191.4	218.4	245.4
2	5.39	32.35	59.32	86.28	113.2	140.2	167.2	194.1	221.1	248.1
3	8.09	35.05	62.01	88.97	115.6	142.9	169.9	196.8	223.8	250.7
4	10.78	37.75	64.71	91.67	118.6	145.6	172.6	199.5	226.5	253.4
5	13.48	40.44	67.41	94.37	121.3	148.3	175.3	202.2	229.2	256.1
6	16.18	43.14	70.10	97.06	124.0	151.0	177.9	204.9	231.9	258.8
7	18.87	45.84	72.80	99.76	126.7	153.7	180.6	207.6	234.6	261.5
8	21.57	48.53	75.49	102.5	129.4	156.4	183.3	210.3	237.3	264.2
9	24.27	51.23	78.19	105.2	132.1	159.1	186.0	213.0	240.0	266.9

220 V, Three-phase, 90% efficiency, 85% power factor

Horsepower	0	10	20	30	40	50	60	70	80	90
0		25.56	51.12	76.68	102.2	127.8	153.4	178.9	204.5	230.3
1	2.56	28.12	53.68	79.24	104.8	130.4	155.9	181.5	207.0	232.6
2	5.11	30.67	56.23	81.79	107.4	132.9	158.5	184.0	209.6	235.2
3	7.67	33.23	58.79	84.34	109.9	135.5	161.0	186.6	212.1	237.7
4	10.22	35.78	60.34	86.91	112.5	138.0	163.6	189.1	214.7	240.3
5	12.78	38.34	64.90	89.46	115.0	140.6	166.1	191.7	217.3	242.8
6	15.34	40.90	66.46	92.02	117.6	143.1	168.7	194.3	219.8	245.4
7	17.89	43.45	69.01	94.57	120.1	145.7	171.3	196.8	222.4	247.9
8	20.45	46.01	71.57	97.13	122.7	148.2	173.8	199.4	224.9	250.5
9	23.00	48.56	74.12	99.68	125.2	150.8	176.4	201.9	227.5	253.0

TABLE 6-17. FULL-LOAD CURRENTS FOR TWO- AND THREE-PHASE MOTORS

Amperes—full-load

Motor horsepower	Squirrel-cage induction											Slip-ring induction										
	Two-phase					Three-phase						Two-phase					Three-phase				
	110-V	220-V	440-V	550-V	2200-V	110-V	220-V	440-V	550-V	2200-V	110-V	220-V	440-V	550-V	2200-V	110-V	220-V	440-V	550-V	2200-V
¼	4.3	2.2	1.1	—	—	5.0	2.5	1.3	1.0	—	—	—	—	—	—	—	—	—	—	—
½	4.7	2.4	1.2	.9	—	5.4	2.8	1.4	1.1	—	—	—	—	—	—	—	—	—	—	—
¾	—	—	—	1.0	—	—	—	—	—	—	6.2	3.1	1.6	1.3	—	7.2	3.6	1.8	1.5	—
1	5.7	2.9	1.4	1.2	—	6.6	3.3	1.7	1.3	—	6.7	3.4	1.7	1.4	—	7.8	3.9	2.0	1.6	—
1½	7.7	4.0	2	1.6	—	9.4	4.7	2.4	2.0	—	11.7	5.9	3.0	2.3	—	—	—	—	—	—
2	10.4	5	3	2.0	—	12.0	6	3.0	2.4	—	12.5	6.3	3.1	2.5	—	14.4	7.2	3.6	2.9	—
3	—	8	4	3.0	—	—	9	4.5	4.0	—	—	8.7	4.3	3.5	—	20.2	10	5.0	4	—
5	—	13	7	6	—	—	15	7.5	6.0	—	—	13.0	6.5	5.2	—	—	15	7.5	6	—
7½	—	19	9	7	—	—	22	11	9.0	—	—	20.0	10.0	7.6	—	—	25	13	10	—
10	—	24	12	10	—	—	27	14	11	—	—	24.3	12.1	10.0	—	—	28	14	11	—
15	—	33	16	13	—	—	38	19	15	—	—	39	19.5	15.6	—	—	45	23	18	—
20	—	45	23	19	—	—	52	26	21	5.7	—	49	24.7	19.8	—	—	56	28	22	—
25	—	55	28	22	6	—	64	32	26	7	—	60	30.0	24.0	6.4	—	67	34	27	7.5
30	—	67	34	27	7	—	77	39	31	8	—	72	36.0	28.8	7.8	—	82	41	33	9
40	—	88	44	35	9	—	101	51	40	10	—	93	46.5	37.3	9.5	—	106	53	42	11
50	—	108	54	43	11	—	125	63	50	13	—	113	57	45	12.1	—	128	64	51	14
60	—	129	65	52	13	—	149	75	60	15	—	135	68	54	14.0	—	150	75	60	16
75	—	156	78	62	16	—	180	90	72	19	—	164	82	65	17.3	—	188	94	75	19
100	—	212	106	85	22	—	246	123	98	25	—	214	108	87	21.7	—	246	123	99	25
125	—	268	134	108	27	—	310	155	124	32	—	267	134	108	27	—	310	155	124	31
150	—	311	155	124	31	—	360	180	144	36	—	315	158	127	32	—	364	182	145	37
175	—	—	—	—	—	—	—	—	—	—	—	—	—	—	—	—	—	—	—	—
200	—	415	208	166	43	—	480	240	195	49	—	430	216	173	44	—	490	245	196	52

E_L = the line voltage between any two phases,

I_L = the line current, and

$\sqrt{3}$ = 1.732.

Current in Three-Phase Circuits

The formula for finding the current in a three-phase circuit can be developed in the same way.

$$I = \frac{P}{E \times pf \times 1.732}$$

Horsepower, Efficiency, and Power Factor in Three-Phase Motors

As in the case of single phase, or in working with DC motors, efficiency is a factor in electrical calculations for three-phase work. If the horsepower, operating voltage, efficiency, and power factor are known, the amount of current can be calculated by

$$I = \frac{746 \times hp}{E \times eff \times pf \times 1.732}$$

This formula is similar to those used previously, except that the $\sqrt{3}$ factor (1.732) is now involved in the calculations.

Example:

How much current is required by a 5-hp, 220-V, three-phase motor whose operating efficiency is 90% and power factor is 80%?

Solution:

$$I = \frac{746 \times hp}{E \times eff \times pf \times 1.732}$$

$$= \frac{746 \times 5}{220 \times 0.9 \times 0.8 \times 1.732} = \frac{746}{44 \times 0.9 \times 0.8 \times 1.732}$$

$$= \frac{746}{54.9} = 13.59 \text{ A}$$

EFFICIENCY

Efficiency, previously considered in connection with DC motors, is also a factor in AC machinery. If the horsepower is known in a single-phase circuit, the current can be calculated by

Efficiency

$$I = \frac{746 \times hp}{E \times eff \times pf}$$

This formula is exactly the same as the formula for current for DC motors. The only difference is that in AC circuits the power factor must be included if it is less than unity.

Example:

An AC motor rated at 3 hp is connected to a single-phase 110-V AC power line. The efficiency of this motor is 90%. What are the current requirements of this motor, rated as having a 90% power factor?

Solution:

$$I = \frac{746 \times hp}{E \times eff \times pf}$$

$$= \frac{746 \times 3}{110 \times 0.9 \times 0.9}$$

$$= \frac{2{,}238}{89.1} = 25.1 \text{ A}$$

Efficiency is also a factor in electrical calculations for three-phase work. If the horsepower, operating voltage, efficiency, and power factor are known, the amount of current can be calculated by

$$I = \frac{746 \times hp}{E \times eff \times pf \times 1.732}$$

This formula is similar to those used previously, except that the $\sqrt{3}$ factor is now involved in the calculations.

Example:

How much current is required by a 5-hp, 220-V, three-phase motor whose operating efficiency is 90% and power factor is 80%?

Solution:

$$I = \frac{746 \times hp}{E \times eff \times pf \times 1.732}$$

$$= \frac{746 \times 5}{220 \times 0.9 \times 0.8 \times 1.732} = \frac{746}{44 \times 0.9 \times 0.8 \times 1.732}$$

$$= \frac{746}{54.9} = 13.59 \text{ A}$$

Example:

A three-phase motor whose efficiency is 90% works with a power factor of 80% from a 208-V power line. It has a 2-hp rating. What is the current drain of this motor?

Solution:

$$I = \frac{746 \times hp}{E \times \text{eff} \times pf \times 1.732} = \frac{746 \times 2}{208 \times 0.9 \times 0.8 \times 1.732}$$

$$= \frac{1{,}492}{259.38} = 5.75 \text{ A}$$

POWER

Power Delivered to a Three-Phase Motor

The formula shown below is applicable to wye and delta wound stator windings.

$$P = \sqrt{3}(E_L)(I_L) \cos \theta$$

where

P = the total power delivered to the three-phase motor,

E_L = the line voltage measured between any two phases,

I_L = the current in a single line, and

θ = the angle between voltage and current in any phase.

Cos θ is the power factor and is equal to R/Z.

Two-Phase Motor Operating from a Three-Wire Line

It is possible for a two-phase motor to work from any two-phases of a three-phase power line. The voltage relationships are expressed by

$$E_d = \sqrt{2}(E_p)$$

where

E_d is the voltage across the stator winding, and

E_p is the voltage measured from the center tap of the stator winding to either side of the line.

Summary of the Characteristics of the Three-Phase Motor

1. Load: Difficult starting load.
2. Starting Torque: 200% to 300% of the running torque.
3. Starting Current: Three to six times as much as the running current.
4. Horsepower: 1/2 to 400 hp.

Terminal Connections for 110-V/200-V Motors

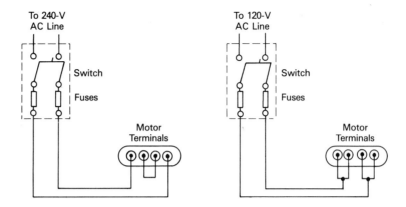

Figure 6-26 Motor terminal connections for 120 V/240 V AC power input.

5. Phase: Three.
6. Source Voltage: 120 to 240 AC, 240 to 480 AC, and higher.
7. Speed: 900, 1,200, 1,800, and 3,600 rpm.
8. Reversibility: Yes.
9. Uses: Industrial machinery, elevators, pumps.

TERMINAL CONNECTIONS FOR 110-V/220-V MOTORS

Some AC motors can operate from either a 120-V or 240-V AC power line. The motor may be equipped with a double-pole, single-throw (DPST) switch for this purpose. Connections for the motor terminal are illustrated in Figure 6-26.

chapter seven

Electric/Electronic Motor Control Circuits

All motors require some form of control, and, depending on the kind and amount of control, can be simple or complex. Motor control includes turning the motor on or off, reversing rotation, stepping the armature through one or more desired angles of rotation, braking, varying torque, controlling horsepower, adjusting the speed, and stopping automatically after a predetermined time. Control can be mechanical or electronic or some combination of the two.

The simplest and possibly the most widely used control is a single-pole, single-throw on-off switch, usually a toggle type. In some instances the switch will be combined with a variable resistor for speed control. An alternative form is a single-pole switch leading to three or more taps on a coil, also for speed control. Instead of a toggle, the switch is a detent type. The arrangement of mechanical controls is usually much simpler, technically, than those involving electronics.

ELECTRONIC POWER SUPPLIES

Battery-operated motors have the convenience of portability, eliminating the need for a trailing power-line cord. Batteries can be made part of the motor assembly, but add to the overall weight. They have the inconvenience of needing either replacement or recharging.

In some instances motors are designed to operate from either battery or electronic power supplies, or may be supplied with either an internal or external charger. In some instances remote control may be required when the motor works in a hazardous environment or where the motor must work at some distance.

Electronic Power Supplies

The electronic power supply for a motor can be the basic half-wave type consisting of nothing more than a single half-wave rectifier but no filter. The output of this supply consists of DC pulses as shown in Figure 7-1(a). Note that for part of the time there is no DC output. It does have the advantage of simplicity and economy. The full-wave power unit in Figure 7-1(b) also consists of pulses, but these come along at twice the frequency of the half-wave.

These two basic supplies can be connected directly to the AC power line, in which case the DC output of the power supplies is determined for the most part by the line voltage. If this is more than required, the voltage can be dropped to the amount required by the motor by a resistor, shown as R. The resistor can be a tapped or variable component. If variable, it may be used to control the DC input to the motor, working as a speed control. In this instance the power supply has a double function: it is used to turn the motor on and off, and it is used to control its speed.

If the pulsating DC supplied by the half-wave and full-wave power supplies affects motor operation adversely, the output of the rectifiers can be followed by a filter consisting of just a single capacitor, often an electrolytic type. The result is a smoother DC output voltage.

Another rectifier type is the full-wave bridge, shown in Figure 7-1(c). Its advantage is that it uses the full output of the power transformer and eliminates the need for an electrical center tap as in the case of the ordinary full-wave supply.

Autotransformer Speed Control

The voltage input to a motor can be controlled through the use of a variable or tapped resistor at the output of the power supply or by the use of an autotransformer as in Figure 7-2. This drawing shows a DC shunt-wound motor whose armature and field windings require different DC voltages. The voltage to the armature is

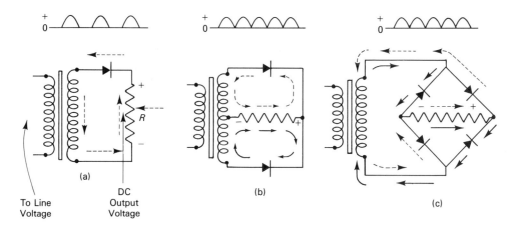

Figure 7-1 Half-wave rectifier (a); full-wave rectifier (b); bridge (c).

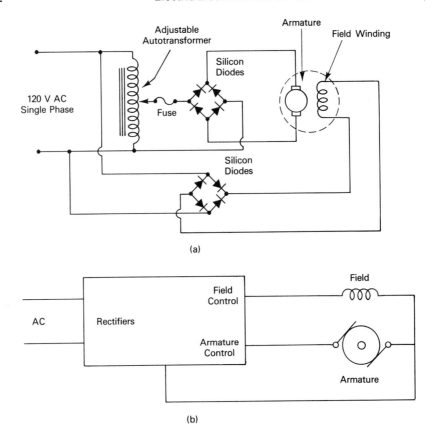

Figure 7-2 Autotransformer used as motor speed controller. Control is obtained by connecting motor to different taps on the transformer (a); overview of electronic circuitry for speed control of DC motors (b).

via a tap on the autotransformer, while the field windings use a bridge rectifier connected directly to the 120-V AC single-phase power line.

Figure 7-2(b) is a block diagram supplying a generalized overview of the technique used for speed control.

Figure 7-3 shows comparable supplies for three-phase power.

Three-Phase Delta-Wye Power

The concept of using an autotransformer for voltage control of a motor can also be used for three-phase line power, as indicated in Figure 7-4. This supply is unusual in that it uses two power-line transformers. One of these is an autotransformer, T1, the other a three-phase transformer, T2. The autotransformer is a variable unit and controls the AC voltage input to T2. T1 is a variable inductor type and

Electronic Power Supplies

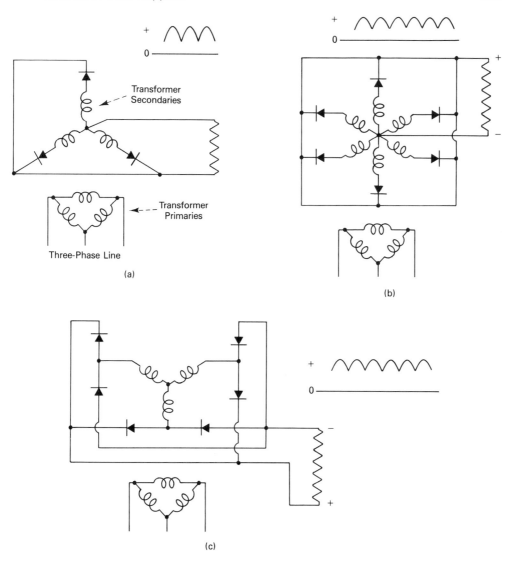

Figure 7-3 Three-phase rectifier systems: half-wave (a); full-wave (b); bridge (c).

so has smooth control of its AC output. Some units are equipped with a calibrated front plate to indicate the amount of voltage output.

The rectifier is a full-wave bridge type with each voltage phase having its own pair of rectifiers. While the rectifiers are not followed by a filter system, the AC input consists of a series of pulses that follow each other rapidly. At no time does the DC voltage drop to zero.

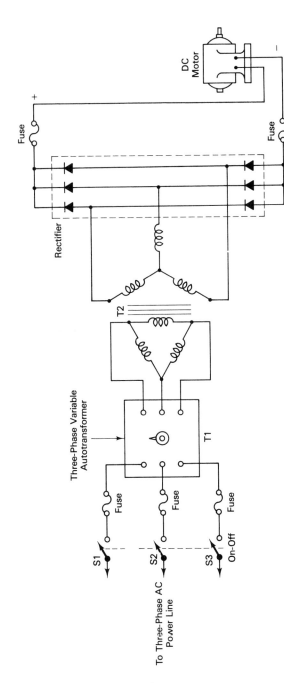

Figure 7-4 Dual-power transformer for three-phase control of a DC motor.

DIAC

A diac is a bidirectional trigger diode and is a member of the thyristor family, a term applied to any semiconductor switch. The diac can be used to trigger a triac into a conducting or nonconducting condition.

TRIAC

A triac is made of a pair of silicon-controlled rectifiers connected in shunt but in reverse. Because of this arrangement a triac can switch direct or alternating currents. The triac has three connections: a gate and a pair of terminals identified as main terminal 1 and main terminal 2, or sometimes simply alphanumerically by A1 and A2. The triac has six semiconductor regions, three n layers plus an n layer sandwiched between two p layers. The gate, and each of the main terminals all contact n-p layers.

When the triac is used to switch alternating current, the unit remains on only when the gate receives operating current. However, if sinusoidal AC is applied, the gate turns off every time the AC waveform passes through its zero point. When a DC voltage is applied to the two main terminals, the unit works like an ordinary silicon-controlled rectifier (SCR).

Like SCRs, triacs are categorized by their current-passing capabilities, and are identified as low-current, medium-current, and high-current types. The current-handling capacity of the triac is in the same range as that of an SCR.

SILICON-CONTROLLED RECTIFIER

Sometimes referred to as a thyristor, the SCR is used for the control of AC or DC power. Like other semiconductor diodes, the SCR is equipped with a cathode and an anode, but in addition it has a third element, a gate. However, unlike the usual diode, it does not conduct in the presence of forward bias until the gate is pulsed. The diode will then continue to conduct until the forward bias is removed or reduced below the conductivity level.

SCR Speed Control

The circuit in Figure 7-5 uses a silicon-controlled rectifier for determining current flow through the armature of a series-wound DC motor. A pair of resistors, R1 and R2, in series, are connected across the rectifiers. R2, a variable resistor, controls the DC input to the gate electrode of the SCR via diode D5. In this way it determines the current flow into the armature, hence its speed. The SCR itself is a rectifier and conducts current in one direction only, and so works both as a rectifier and a control device. Also known as a unidirectional thyristor, its on and off times are measured in microseconds.

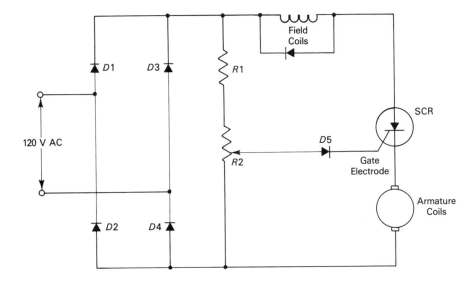

Figure 7-5 Electronic speed control of a series-wound DC motor.

The power supply uses four diodes, D1 through D4, in a full-wave bridge circuit. The circuit is simplified by the absence of a power transformer and a filter. Since the field and armature windings are in series, the SCR controls current flow through both.

CHOPPER ACTION

The delivery of a DC voltage to a motor is either from a battery and is smooth and continuous, or is from a non-filtered electronic power supply and is pulsating. In either case, the use of a commutator breaks the DC into a series of rapidly occurring pulses. Instead of a mechanical commutator, an electronic type called a *chopper* can be used instead. In the case of an electronic chopper, the frequency of the pulses can be controlled by the circuitry and thus is independent of the motor.

SPEED CONTROL OF BATTERY-OPERATED SERIES-WOUND DC MOTORS

There are various ways of controlling the speed of a battery-operated series-wound type of motor, and all of them accomplish this by voltage adjustment.

The first and easiest way is to put a variable resistor in series with the DC supply to the motor. This puts the resistor in the same line as the current flow to the field and armature coils. The resistor must be able to carry this current. The power developed in the resistor is $P = I^2R$ and is wasted power in the form of heat. It has the advantage of simplicity.

Switching is another technique and consists of putting more batteries in series with the main voltage supply, but this isn't very practical. It can be done electronically with the circuit shown in Figure 7-6. Known as a capacitor voltage boosting circuit it consists of an inductor shunted by a capacitor. When the switch, S1, is in position 1, a current flows through the coil, L, building a magnetic field around it. When the switch is in position 2, the magnetic field collapses, producing a voltage across the capacitor. This is in addition to the voltage supplied by the battery, so the total EMF is greater than that of the battery alone. S1 is shown as a mechanical switch, but an electronic type is used.

The final method involves using a chopper circuit in series with the motor. The chopper is simply a rapidly acting electronic switch. Figure 7-7 shows the waveform output of a chopper. During its off time the DC output has a very low average. This becomes higher during the time the chopper is on. Motor speed can be controlled by adjusting the ratio of on to off time of the chopper.

SPEED CONTROL FOR UNIVERSAL MOTOR

Since the universal motor can operate from the AC line, the power supply can be omitted, as in Figure 7-8. The motor is in series with the SCR, with these two units connected to the AC single-phase source. The amount of current flow through the motor is controlled by the SCR. This component, in turn, has its current flow determined by the setting of variable resistor R2. The flow of current through R2 is pulsating DC. R3 and C2 work as a filter and help maintain the voltage supplied by R2 at a constant amount.

DIRECTION OF ROTATION OF AC MOTORS

The method used for changing the direction of rotation of AC motors depends on the motor type.

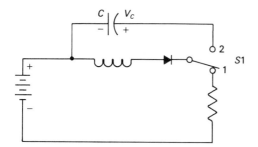

Figure 7-6 Voltage booster.

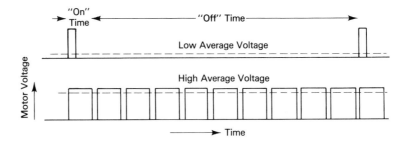

Figure 7-7 Average output voltage of chopper depends on ratio of on time to off time.

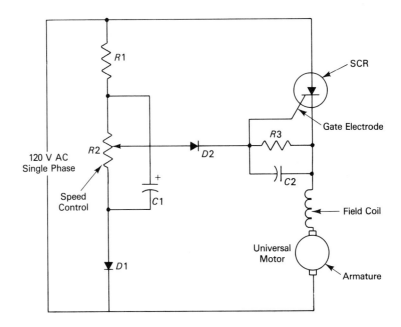

Figure 7-8 Electronic speed control for universal motor.

Single-Phase Capacitor Motor

As shown in Figure 7-9, the single-phase capacitor motor consists of a pair of series-connected main windings plus a pair of auxiliary windings, also wired in series. These are connected to the source voltage. The capacitor is used to obtain a phase shift for starting the motor.

Change of direction is obtained by a toggle switch, a single-pole, double-throw type. This switch is identified by the abbreviations "for" (forward) and "rev" (reverse). The change of armature direction is via the capacitor, which is switched from one main winding to the other.

Direction of Rotation of AC Motors

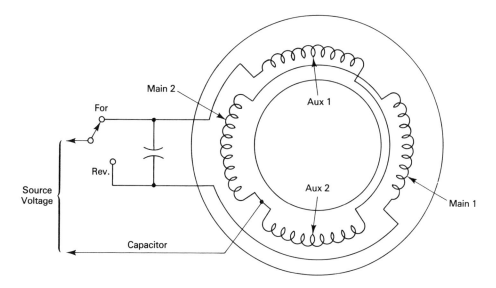

Figure 7-9 Technique for reversal of single-phase capacitor motor.

Single-Phase Series Motor

Some single-phase series motors can have the direction of rotation of the armature changed by moving the brushes from one side of the neutral plane to the other. Some motors may have the position of the neutral plane marked and the brush assembly (consisting of the brush and its holder) designed for easy repositioning so as to minimize commutator sparking or to reverse armature direction (Figure 7-10(a)).

Split-Phase Motors

Split-phase motors use a special inductor, a starting winding, to obtain initial armature rotation. The starting winding may be disconnected by a centrifugal switch after the motor has reached about 80% of its running speed. The starting winding consists of many turns of fine wire; the running wire of fewer turns having a lower-gauge-number. To reverse the running direction, transpose the connections of either the starting winding or the running winding, but do not transpose the leads to both coils (Figure 7-10(b)).

Two-Phase Motors

Transpose the connections of either one of the phase windings to reverse the direction of rotation.

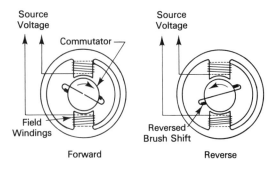

Figure 7-10(a) Change of direction of armature rotation may be obtained by moving the position of the brushes.

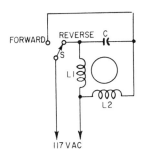

Figure 7-10(b) Method of reversing rotation of split-phase induction motor.

FRACTIONAL-HORSEPOWER MOTORS

Some motors are designed to run in one direction only and so may be categorized as non-reversible. This does not mean it is impossible to run them in the reverse direction. Shaded-pole motors are usually built to turn just one way, but some can be reversed by transposing their shading rings, as indicated in Figure 7-11.

Some motors can be reversed only when the armature is at rest; others can be reversed with the armature running at its rated speed. This can be done by reversing the direction of current flow through the armature windings or through the field windings of motors that use a commutator and a brush assembly. Induction-type motors can be reversed by changing the direction of rotation of the magnetic field.

The operating characteristics of a motor may be changed by reversing. Motors that must be stopped for reversal maintain their operating specifications; those that are reversible during running are affected. Usually there is some sort of trade-off in characteristics.

A shunt-wound motor may take advantage of a skewed-slot armature so as to improve its reversing torque. The greater the angle of skew, the faster the reversal, but this is achieved at the expense of speed regulation. If the magnetic field is strong but the air gap between the armature and field is larger, armature reversal is improved, but the running efficiency is reduced. The chart in Table

Fractional-Horsepower Motors

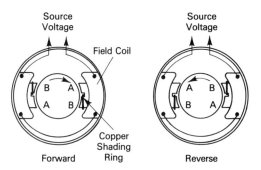

Figure 7-11 Direction reversal by transposition of shading rings.

7-1 indicates the features of fractional-horsepower motors having a reversing capability.

Reversing the DC Shunt-Wound Motor

The DC shunt-wound motor can be reversed by changing the direction of current flow to either the shunt field or the armature. Figure 7-12 shows how this can be done by using a four-pole double-throw switch inserted in the connecting leads to the armature. One of the advantages of reversing the armature current instead of the field is that it tends to equalize the wear of both sets of brushes, thus increasing brush life.

Reversing the DC Compound-Wound Motor

The DC compound-wound motor type is reversed by changing the direction of current flow through the armature winding. The shunt and series fields do not have their direction of flow changed. The reversing characteristics of this motor type are generally better than those of the shunt wound, since the series coil increases its field strength during the reversal time. While this improves the reversing torque, it does lower the speed regulation (Figure 7-13).

Reversing the Series-Wound Motor

Rotation can be reversed in the series-wound motor by changing the direction of current flow through either the field coil or the armature winding. This is illustrated in Figure 7-14.

Reversing Polyphase Induction Motors

Reversing a three-phase motor can be done by transposing any two of its connections to the power line. Polyphase motors have good starting and reversing characteristics. Whether using a delta or wye arrangement, the result is a balanced rotating electromagnetic field.

TABLE 7-1. FRACTIONAL-HORSEPOWER REVERSING MOTORS

Fractional-horsepower motors from 1/2000 hp to 1/4 hp	Features of fractional-horsepower reversing motors													
	Induction-type reversing motor								Brush-type reversing motor				
	Split phase		Capacitor start and run				Multiphase		Shunt motor		Compound motor	Series motor	
			Non-synch.		Synch.								
	Non-synch.	Synch.	4-Lead rev.	3-Lead rev.	4-Lead rev.	3-Lead rev.	Non-synch.	Synch.	Full field	Split-field	5-Lead reversible	4-Lead rev.	3-Lead rev.
Suitable for reversing at rest only	X	X											
Suitable for reversing during rotation or at rest			X	X	X	X	X	X	X	X	X	X	X
Double-pole double-throw switch required		X	X		X				X		X	X	
Single-pole double-throw switch sufficient				X		X				X			X
Running characteristics slightly affected by best obtainable instantaneous reversing characteristics			X		X		X	X	X		X	X	
Running characteristics greatly affected by best obtainable instantaneous reversing characteristics				X		X				X			X

Speed and Regulation Control

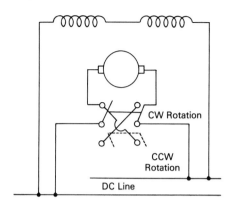

Figure 7-12 Four-wire reversible shunt-wound motor.

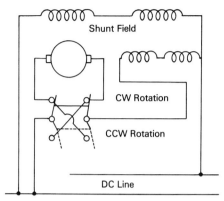

Figure 7-13 Reversing technique for five-wire compound motor.

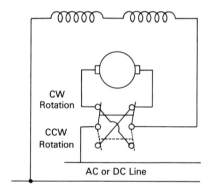

Figure 7-14 Reversing method for series-wound motor.

SPEED AND REGULATION CONTROL

In some instances an electronic control circuit will be able to perform more than one function. Such a circuit is shown in Figure 7-15. It is intended to work both as a speed and as a regulation device. Regulation is a reference to a motor's variation in speed with changes in its load. A motor has good regulation if it maintains its speed despite load increases and decreases.

Shunted across the 120-V single-phase AC power line is a series network consisting of R1, variable resistor P1 (the speed control), and a diode rectifier, D1. Because of the presence of D1, the current flowing through the speed control is DC. Since D1 is a half-wave rectifier, P1's voltage would ordinarily consist of a series of pulses. But C2 works as a filter, smoothing P1's output. The capacitor is a high-capacitance electrolytic. Additional filtering is supplied by R4 and C4. P1 determines the operating condition of transistor Q1, since it supplies it with DC bias. This transistor works as an emitter-follower and supplies the gating voltage

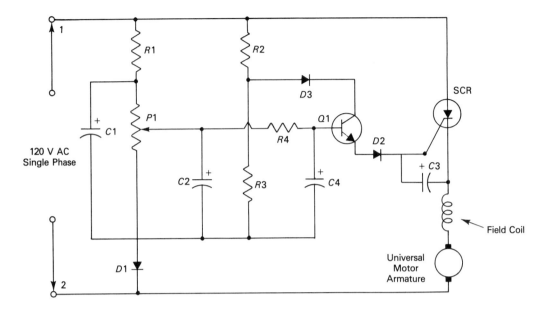

Figure 7-15 Speed and regulation control of a universal motor.

for the SCR. While P1 determines the operating speed of the motor, the speed will remain fairly constant under varying load conditions.

The SCR is in series with the field coil and armature of the universal motor; thus these three components will carry the same current. The SCR will trigger into conduction under these conditions: its gate drive input must be equal to the counter EMF of the motor plus the triggering voltage it receives from P1 via Q1. Of these two voltage inputs, that supplied by Q1 is the reference voltage.

If the motor speed decreases, so will the counter EMF. This lets the SCR trigger on with a lower drive voltage. With the SCR conducting, the motor voltage rises and so does the motor speed. If the motor speed increases, an opposite effect takes place.

Motor Braking

When a motor is used in an application requiring changes in motor speed, as in the case of an electric drill, braking can be accomplished by the insertion of resistance in the line. The electric drill uses a trigger for variable resistance control.

In some instances the source voltage is removed and a brake resistance is inserted in series with the armature, as in the DC shunt-wound motor of Figure 7-16(a). A similar method is used for the DC compound wound motor in Figure 7-16(b). An alternative method for the series-wound DC motor is to insert a brake

Speed and Regulation Control

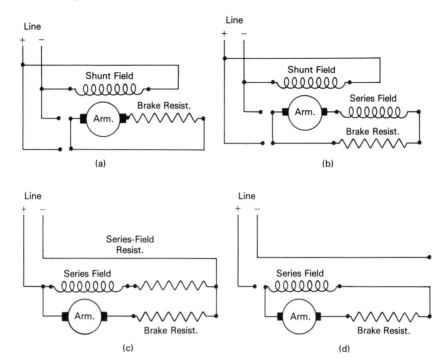

Figure 7-16 Braking methods for shunt-wound (a) and compound-wound (b) motors; two methods of braking for series-wound motors (c and d).

resistance without disconnection from the DC line as in Figure 7-16(c). However, brake disconnection from this motor is also used, as in Figure 7-16(d).

Diac and Triac Speed Control

A diac is a bidirectional diode thyristor, a semiconductor switch. A triac is a bidirectional triode thyristor. The circuit in Figure 7-17 uses a diac and a triac for the speed control of a universal motor.

The AC drive voltage can be either 120 V AC or 240 V AC, with both voltages single phase 60 Hz. For 120 V, R1 is 100 kΩ, 1/2-W, and for 240 V it is 200 kΩ, 1 W. For 120 V AC, C1 is 0.22 μF, 200 working volts DC (WVDC) with the working voltage increased to 400 for 240-V input.

R1 and C1 form a phase-shifting network, which is put across the AC input in series with the motor windings. Variable resistor R1 is the speed control. R2 and C2 work as a voltage divider, with the diac connected to their meeting point.

The triac is equivalent to a pair of SCRs connected back-to-back. R3 and C3 are wired in series and are shunted across the triac. They work as a snubber and are used to absorb the inductive kickback pulse that can develop when the motor is switched off.

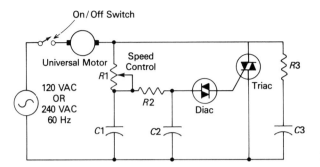

Figure 7-17 Diac and triac speed control.

The triac can be regarded as a variable resistor controlling the amount of current supplied to the windings of the universal motor.

ELIMINATION OF TRANSIENT VOLTAGES

With the use of a power supply a DC motor can work from the AC line instead of depending on batteries. The DC motor may have wanted characteristics, but the use of a battery to supply the required amount of motor current might be impractical.

One of the problems of using the AC line is that it is sometimes accompanied by high peak transient voltages that could change motor speed at a time when the motor speed should remain constant.

Figure 7-18 shows the circuit for a shunt wound DC motor control also equipped with transient voltage control. Shunted across the AC line input is a thyrector used for eliminating or minimizing the amplitude of transients. While this circuit is connected directly to the power line, a 1:1 power transformer could also be used to provide isolation from other components plugged in to the same line.

The power supply is a full-wave bridge, with the DC output applied directly to the field coils. That same voltage is also used for the armature connected in series with an SCR.

The SCR is controlled by two factors: the time it takes capacitor C1 to charge to the breakdown voltage of the diac and the breakdown level characteristic of that diac. Resistor R2 is the speed control.

RFI FILTER

Motors, especially those using brushes and a commutator, are capable of causing severe AM radio and television interference. This interference can be transmitted by the use of a common power line or can be radiated by the motor in the manner of a broadcast signal.

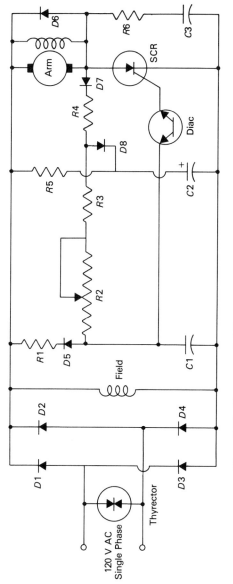

Figure 7-18 Shunt-wound DC motor speed control.

The arrangement in Figure 7-19 is a phase control circuit for a universal motor speed control. It includes a radio-frequency-interference (RFI) filter for the suppression of electrical noise. R1 is the speed control, with its output voltage triggering the diac. Resistor R3 and capacitors C1 and C2 act as an R-C filter for this voltage, making certain that any pulse voltage from R1 will not trigger the diac accidentally.

Table 7-2 supplies information about C1, C2, R1, R2, and R3 for two different AC inputs: 120 V at 60 Hz and 240 V at 50 Hz. Some experimentation may be required with the values indicated for the RFI filter.

Induction Motor Control

The circuit diagram in Figure 7-20 is intended for certain types of induction motors such as shaded-pole or permanent split-capacitor types operating with fixed loads. The circuit is intended for speed control for fans or blowers, where a small change in armature speed results in a large change in air velocity. In some instances, as

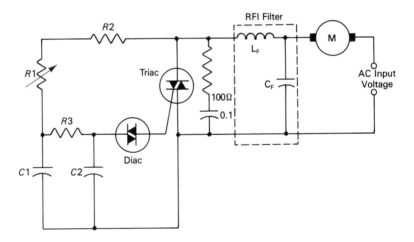

Figure 7-19 Phase control circuit for governing speed of universal motors.

TABLE 7-2. COMPONENT VALUES FOR FIGURE 7-19 FOR 120 AND 240 V

AC input voltage	C1	C2	R1	R2	R3	RFI filter	
						L_F	C_F
120 V 60 Hz	0.1 μF 200 V	0.1 μF 100 V	100 kΩ 1/2 W	2.2 kΩ 1/2 W	15 kΩ 1/2 W	100 μH	0.1 μF 200 V
240 V 50 Hz	0.1 μF 400 V	0.1 μF 100 V	250 kΩ 1 W	3.3 kΩ 1/2 W	15 kΩ 1/2 W	200 μH	0.1 μF 400 V

Torque Control

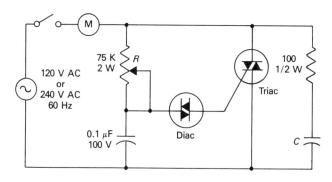

Figure 7-20 Speed control for induction motors.

in the case of induction motors, the motor may stall if the speed of rotation is reduced below its dropout speed.

The potentiometer R is the speed control. It determines the conduction point of the diac. The value of capacitor C is determined by the AC voltage input. For 120 V AC, 60 Hz, it is 0.22 µF, 220 V, while for 240 V AC, 60 Hz, it has the same capacitance but a higher working voltage, 400 V.

Motor-Reversing Circuitry

The circuit in Figure 7-21 uses a pair of triacs to obtain a motor-reversing capability. Reversing can be had manually by operating a switch or electronically. A very low-value resistor, 2 Ω, is shown in series with a capacitor. This network is used to limit the discharge current of the capacitor to a safe value when both triacs conduct at the same time.

TORQUE CONTROL

Figure 7-22 is a block diagram of a circuit used for torque control. The unit, a power converter, controls torque but does so without requiring an increase in armature speed. As a result the armature current remains constant.

The torque of a motor depends on the power developed by a rotating armature. Power equals $E \times I$. If more torque is needed and if the current is to remain constant, the alternative is to increase the voltage supplied to the armature.

In the diagram resistor R1 is in series with the armature and so carries the full armature current. The voltage across this resistor, E, is equal to $I \times R$. This voltage is fed back to an integrated circuit, which, in turn, triggers firing circuits to supply more voltage to the armature when the armature current decreases. The opposite effect takes place when the armature current increases.

Usually, an increase in the load results in a rise in armature current. However, the feedback circuit works to raise the armature voltage. The current, then, moves back to its original level.

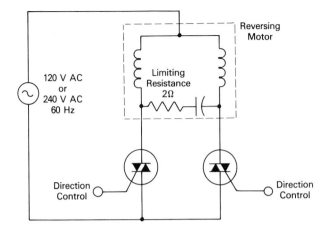

Figure 7-21 Direction control for reversing motors.

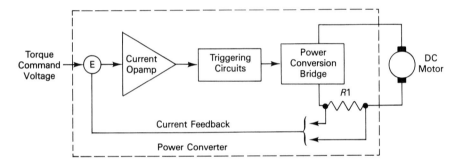

Figure 7-22 Circuit for torque control.

HIGH-CURRENT, LOW-SPEED VOLTAGE REGULATOR

The drawing in Figure 7-23 is a combination block and circuit diagram. The DC motor requires an input between 12 and 20 V and has a running current between 1 and 5 A. The motor's speed is adjusted by a 25kΩ variable resistor with this resistor shunted across an adjustable voltage regulator. The amount of output voltage is proportional to the ratio of R2 to R1.

WHEATSTONE BRIDGE CONTROL

The resistor arrangement in Figure 7-24(a), consisting of R1, R2, R3, and R4, is known as a Wheatstone bridge. If the resistors have equal values of resistance and if they have the same amount of current flow, then each of the resistors will have the same amount of voltage across it. When this happens, the voltage measured from point A to point B will be zero.

Tachometer Motor Control

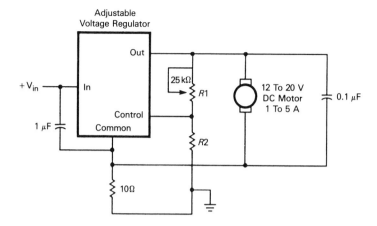

Figure 7-23 High-current, low-voltage speed regulator.

As shown in Figure 7-24(a), an amplifier is connected across these two points. When the voltage between points A and B is zero, there is no voltage input to the amplifier, thus the field winding of the motor receives no current. The motor remains inoperative.

The drawing in Figure 7-24(b) is almost identical with that of Figure 7-24(a) except that the motor shaft is connected to a variable resistor identified as R. This resistor is a heating element for some industrial process with the current through it supplied by a DC voltage source. Although they are shown as two separate resistors, R and R4 are the same. R4 is a temperature-sensitive element. The objective of this circuit is to have the temperature of R4 remain as constant as possible.

If the temperature of R4 should change, so will its resistance. The current flow through it will also change, producing a voltage difference between points A and B. This voltage will be supplied to the amplifier, and a current will be delivered to the field coils. The motor shaft will turn in such a direction that the value of resistance will now change in an opposite direction. This action will continue until the value of resistor R (actually R4) will be the same as that across the other resistors in the bridge. At that time there will be no voltage across points A and B, no input to the amplifier, and the motor will stop running.

TACHOMETER MOTOR CONTROL

A tachometer is used for measuring the speed of rotation of a motor shaft and can be used for motor control for both AC and DC motors. The tachometer is a voltage generator, is coupled physically to the motor shaft, and produces a voltage that is proportional to the speed of the motor.

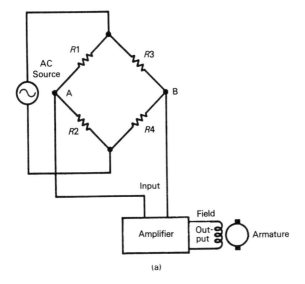

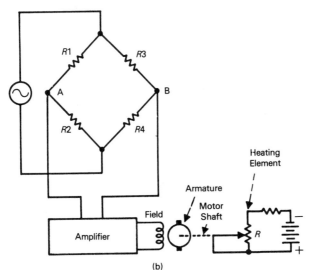

Figure 7-24 Wheatstone bridge motor control (a); motor-operated temperature control (b).

As indicated in Figure 7-25 the output voltage of the tachometer is fed back to an error operational amplifier integrated circuit and is delivered to a comparator circuit identified as E. At the same time, the motor to be controlled (a DC type in this example) supplies current feedback to the comparator. The current is converted to a voltage by sending it through resistor, R.

The two voltages, that from the speed command and that from the motor, are out of phase. As long as a difference voltage, plus or minus, exists, it is fed to the following operational amplifier and from there to a voltage amplifier and then

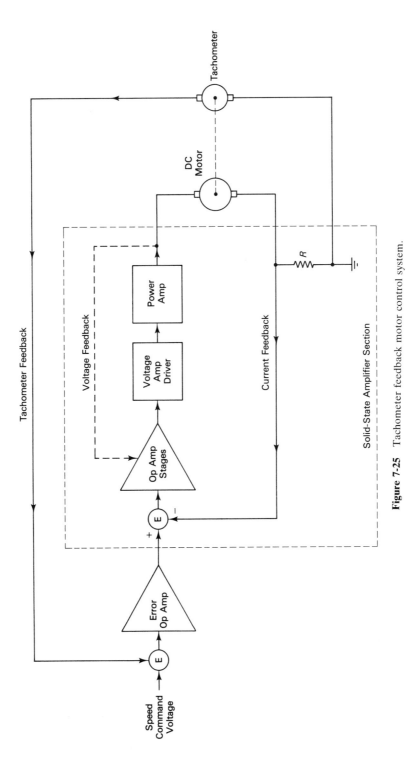

Figure 7-25 Tachometer feedback motor control system.

to a power amplifier, driving the motor in one direction or the other. This direction is such that the current feedback (in voltage form) and the error voltage from the operational amplifier (op amp) cancel. The desired speed is controlled by the speed command voltage control. Once it is set the motor will continue to change its speed until the wanted speed is reached.

There are a number of circuit variations of this speed control technique, but in general they follow the feedback error-correction system shown in Figure 7-26. The controlled motor returns a positive or negative voltage an error-measuring device. In this circuit a comparison is made between the required speed and the actual speed, in terms of voltages. The difference between the two is fed into an amplifier and then into the field windings of the motor. Once the speed of the motor reaches its predetermined amount, no further error voltage is developed and the motor continues running at its required speed. If that speed should change, either increasing or decreasing, a correction error voltage would be developed once again, and the speed correction process would take place.

HIGH-FREQUENCY MOTOR CONTROL

Land-based AC motors use a 50-Hz or 60-Hz power line. Planes and ships are equipped with their own AC power generators, and these may have an output of 115 or 208 V RMS. The waveform is sinusoidal, and the operating frequency is often 400 Hz.

The motor speed control (Figure 7-27) is a 100-kΩ, 1/2-W variable resistor, R1. When the triac becomes conductive, its current demand increases the current flow through the load, a motor in this example. This increased flow raises the motor speed.

R1, R2, and C1 form a phase-shifting network. The greater the resistance of R1 and R2 in series compared to the capacitance of C1, the smaller the phase angle. When R1 is in its minimum-resistance position, the phase angle is larger. It is this voltage phase angle that controls the conductivity of the triac.

The series resistor, R3, and capacitor, C2, enclosed in dashed lines, form a snubber. When the current to the field coils and the armature coils is cut off abruptly, the extremely rapid collapse of their magnetic fields induces a very high

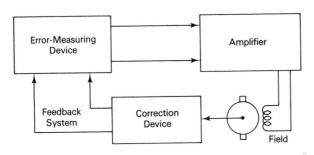

Figure 7-26 Feedback error-correction system.

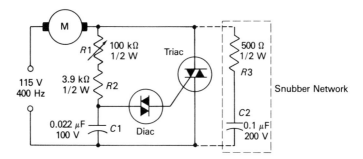

Figure 7-27 High-frequency AC input using motor control.

voltage across the coils. R3 and C2 absorb the energy of the swift change in magnetic field strength, preventing the strong induced EMF from damaging circuit components.

SERIES MOTOR SPEED CONTROL

The circuit in Figure 7-28 is a simple type of speed control for a series-wound motor, but it does have the disadvantage of requiring that separate connections should be available for the field and armature coils. While the motor is a series-wound DC type, it gets its source voltage from a full-wave power supply connected to the 120-V AC line. The power supply is unfiltered, but an electrolytic capacitor could be connected across the power supply to furnish a smoother output. However, the regulation of the power supply would be decreased. The field windings, the SCR and the armature coil are all in series, so the same current flows through all

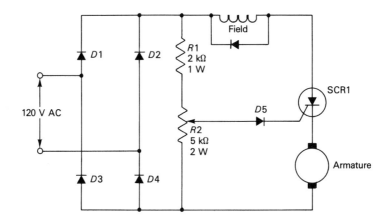

Figure 7-28 Speed control of series-wound DC motor.

of them. Control of this current is via SCR1, with its conductivity adjusted by the 2-W, 5-kΩ variable resistor, R2.

Resistor R2 controls the motor speed. This happens when the voltage supplied by this resistor exceeds the voltage drop across diode D5 and the voltage drop from the gate to the cathode of SCR1. Under these conditions the armature will increase its speed, increasing its counter EMF at the same time. The speed of the motor then adjusts to the voltage supplied by R2. One of the problems of this circuit is that the motor may hunt for various setting of R2.

SENSOR-SERVO CONTROL

The block diagram in Figure 7-29 is that of a feedback system using a sensor and a servo. The feedback system, shown as a potentiometer, is a device that works as a transducer, changing the speed of the motor to a corresponding voltage.

The motor control potentiometer, shown at the left, is set to supply the initial operating speed of the motor. There are two voltage inputs to the sensor, one from the control potentiometer and another from the feedback potentiometer. These voltages are out of phase, and if they have identical amplitudes, there is no signal output from the sensor. If the motor should increase or decrease its speed from the established norm, the output from the feedback potentiometer, known as an *error voltage*, will increase or decrease. This, in turn, will supply a sensor output voltage that is supplied to the following servo.

The servo amplifier will go into action to speed or slow the motor until it operates at the predetermined speed. When it reaches this speed, the two feedback voltages once again have the same amplitudes, resulting in zero input to the servo.

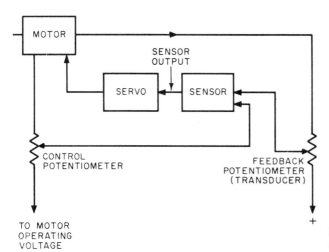

Figure 7-29 Motor control using sensor-servo feedback system.

Micromotors

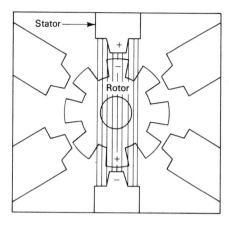

Figure 7-30 Action of a micromotor.

MICROMOTORS

Micromotors are so small in dimension, practically invisible, that two of them could be positioned side by side across the width of a human hair. A silicon chip, possibly 1/4" square, could accommodate a number of them easily. The rotor measures about 2 milli-inch.

Ordinary motors take advantage of the forces of attraction and replusion between magnets. The micromotor is operated, instead, by electrostatic forces, the same that are at work when hair clings to a comb or the spark that occurs when you walk across a rug and then touch a metal door knob.

In the micromotor (Figure 7-30), a number of stators are made to surround a freely revolving rotor. A positive voltage is applied to the uppermost stator; a negative voltage to its opposite and lowermost stator. No current flows between these two poles since they are not physically connected but lines of electrostatic force exist between them. These lines of force induce an elestrostatic voltage in two of the adjacent rotor elements, as indicated in the illustration.

The voltage that was initally applied is then removed from the two stator poles and shifted to the next adjacent pair, either clockwise or counterclockwise. As a result the induced electrostatic voltage in the stator also shifts. But in making this shift the induced voltage forces the rotor to turn.

This action is comparable to the rotating magnetic field of motors that use such fields to achieve movement of an armature. The speed of the electrostatically induced voltages controls the speed of the rotor and that is based on the speed of rotation of the stator's electrostatic voltages.

chapter eight

Solved Motor Problems

Often enough the solution to a motor problem involves nothing more than selecting the right formula, plugging in the data, and doing the necessary arithmetic. However, problems can have ascending orders of difficulty, including conversions from multiples and submultiples to basic units, manipulation of a formula, and the use of two or more formulas. In some cases the data may be obscured by the way in which the problem is presented. The purpose of this chapter is to supply a sufficient number of solved problems, ranging from simple to complex, to provide familiarity with the techniques.

Example:

A series resistive network in a motor control circuit consists of three resistors: 860 Ω; 1.2 kΩ; and 0.06 MΩ. What is the total resistance?

Solution:

$$R_t = R1 + R2 + R3$$

$$1.2 \text{ k}\Omega = 1{,}200 \text{ }\Omega$$

$$0.06 \text{ M}\Omega = 60{,}000 \text{ }\Omega$$

$$R_t = 860 + 1{,}200 + 60{,}000 = 62{,}060 \text{ }\Omega$$

Example:

A series-parallel resistor network consists of $R1 = 10$ Ω and $R2 = 24$ Ω, which are connected in parallel. These resistors are in series with $R3$ whose value is 33 Ω. The total current is 2 A. What is the total resistance and the voltage across each resistor?

Solved Motor Problems

Solution:

$$R_t = \frac{R1 \times R2}{R1 + R2} + R3$$

Combine the 10-Ω and 24-Ω resistors:

$$R_t = \frac{R1 \times R2}{R1 + R2} = \frac{240}{34} = 7.06\ \Omega$$

Combine the series and parallel network:

$$R_t = 7\ \Omega + 33\ \Omega = 40\ \Omega$$

The voltage across $R3$ is

$$E = I \times R = 2 \times 30 = 60\ \text{V}$$

The total current is 2 amperes and this is the amount of current flowing through $R1$ and $R2$. The equivalent resistance is 7.06 ohms. $E = I \times R = 2 \times 7.06 = 14.12$ volts, and is the voltage across the parallel resistors. The total voltage $= 14.12 + 60 = 74.12$ volts.

Example:

The name plate on a DC motor indicates that its rating is 3 hp and that it is to be connected to a 100-V source. The efficiency of this motor when supplying its rated horsepower output is 80%. What is the line current supplied to this motor?

Solution: Convert 3 hp to watts: 3 hp × 746 = 2,238 W. Calculate the amount of line current at 100% efficiency:

$$I = \frac{P}{E} = \frac{2{,}238}{100} = 22.38\ \text{A}$$

At an efficiency of 80% the line current is

$$I = \frac{22.38}{0.80} = 27.98\ \text{A}$$

Note that the lower the efficiency the greater the required amount of line current.

Example:

A 1-hp DC motor operates with an efficiency of 75%. The line voltage supplied by a motor-generator is 110 V. What is the current input to this motor?

Solution:

$$I = \frac{746 \times hp}{E \times \text{eff}}$$

$$= \frac{746 \times 1}{110 \times 0.75} = 9\ \text{A}$$

Example:

What is the horsepower output of a 110-V DC motor having a current of 5 A and an efficiency of 85%?

Solution:

$$hp = \frac{E \times I \times \mathit{eff}}{746}$$

$$= \frac{110 \times 5 \times 0.85}{746} = 0.6267 \text{ hp}$$

Example:

A series resistor connected to the input of a series-wound DC motor dissipates 25 W of power and has a voltage drop of 15 V. How much current flows through the armature of this motor?

Solution: Since this resistor is in series with the armature winding, the same amount of current flows through both.

$$I = \frac{P}{E} = \frac{25}{15} = 1.667 \text{ A}$$

Example:

A resistive load of 300 Ω is connected to a DC voltage source measured at 220 V. What is the power utilized by this resistor?

Solution:

$$P = \frac{E^2}{R} = \frac{E \times E}{R} = \frac{220 \times 220}{300} = 161.3 \text{ W}$$

Example:

What is the efficiency of a 5-hp DC motor whose input power is 5.2 kW with the motor operating at full load?

Solution:

$$\mathit{eff} = \frac{\text{output}}{\text{input}}$$

$$5 \text{ hp} = 5 \times 746 = 3{,}730 \text{ W} \qquad 5.2 \text{ kW} = 5{,}200 \text{ W}$$

$$\mathit{eff} = \frac{3{,}730}{5{,}200} = 0.7173 = 71.73\%$$

Example:

Three cells are connected in series aiding. These cells have different terminal voltages: 1.47, 1.38, and 1.30 V, respectively. What is their total voltage? The internal resistance of each of the cells is 0.15, 0.16, and 0.11 Ω. What is the total internal resistance? When a load is connected across the output of these series-wired cells, the amount of current flow is 385 mA. What is the voltage output of the cells under load?

Solution: The total voltage under conditions of no load is

$$E_t = E1 + E2 + E3 = 1.47 + 1.38 + 1.30 = 4.15 \text{ V}$$

Solved Motor Problems

The total internal resistance of the cells is

$$R_t = R1 + R2 + R3 = 0.15 + 0.16 + 0.11 = 0.42 \ \Omega$$

The internal voltage drop is

$$E = I \times R = 0.385 \times 0.42 = 0.1617 \text{ V}$$

The total output voltage under load is

$$E_{output} = E_{no\ load} - E_{full\ load}$$
$$= 4.15 - 0.1617 = 3.9883 \text{ V}$$

Example:

The diameter of a solid wire is 210 mils. What is the cross-sectional area of this wire in circular mils?

Solution:

$$A = d^2 = 210 \times 210 = 44{,}100 \text{ circular mils}$$

Example:

The dimension of one side at the end of a square bar magnet is 5 cm. It is estimated that there are 2.10×10^6 lines of flux (maxwells) existing at this end of the magnet. What is the flux density in gauss?

Solution:

$$B = \frac{\phi}{A} = \frac{2{,}100{,}000}{5 \times 5} = 84{,}000 \text{ gauss}$$

Example:

The cold resistance of the copper wound field coils of a DC motor is 510 Ω when measured at 25°C. After 3 hr of working under load, their temperature rises to 60°C. What is the resistance of the field windings under these conditions?

Solution:

$$\frac{R1}{234.5 + T1} = \frac{R2}{234.5 + T2}$$

By substituting the data supplied in the problem the proportion becomes

$$\frac{510}{234.5 + 25} = \frac{R2}{234.5 + 60}$$

Cross-multiplying supplies

$$259.5 \ R2 = 510 \times 294.5$$

$$R2 = \frac{510 \times 294.5}{259.5} = 578.79 \ \Omega$$

Example:

A pair of copper wires, each having a total length of 500 ft and a specific resistivity of 10.37 Ω at 20°C are used to supply power to a DC motor. The wires have a cross-sectional area of 5,500 circular mils. What is its resistance at this temperature?

Solution:

$$R = \frac{10.37 \times 1{,}000}{5{,}500} = 1.885 \ \Omega$$

Example:

A DC motor has an input of 800 W and an output of 3/4 hp. What is the efficiency of this motor?

Solution:

$$eff = \frac{\text{output}}{\text{input}}$$

$$3/4 \text{ hp} = 3/4 \times 746 = 0.75 \times 746 = 559.5 \text{ W}$$

$$eff = \frac{559.5}{800} = 0.699 = 69.9\%$$

Example:

A single-phase AC motor rated at 5 hp operated from a 120-V power line. At its full rated load its efficiency is 90%. It has an 85% power factor. What is the line current going into this motor?

Solution:

$$5 \text{ hp} = 5 \times 746 \text{ W} = 3{,}730 \text{ W}$$

$$\text{Input Power} = \frac{\text{Output Power}}{eff}$$

$$= \frac{3730}{0.85} = 4388.235 \text{ W}$$

To determine the line current,

$$I = \frac{P}{E \times pf}$$

$$= \frac{4388.235}{120 \times 0.85} = 43.022 \text{ A}$$

Example:

A freight elevator is built to lift a weight of 2,000 lb a distance of 90 ft in 25 s. The efficiency of the machinery is 75%, while the efficiency of its drive motor is 80%. What is the horsepower input to the elevator machinery?

Solved Motor Problems

Solution: The overall efficiency of this system is the product of the motor's efficiency and that of the elevator machinery.

$$\text{Overall efficiency} = \text{machinery efficiency} \times \text{motor efficiency}$$
$$= 0.75 \times 0.80 = 0.60 = 60\%$$

Calculate the output power:

$$P_o = \frac{F \times d}{33{,}000 \times t}$$

(F is in pounds, d is in feet, and t is in minutes.)

$$= \frac{2000 \times 90}{33{,}000 \times 25/60} = 13.1 \text{ hp}$$

$$P_i = \frac{P_o}{\text{eff}} = \frac{13.1}{0.60} = 21.83 \text{ hp}$$

Example:

A DC motor designed for an input of 110 V is located 300 ft from its source. The amount of current flow through the connecting wires is 3 A, and the wire has a diameter of 1/8 in. What is the amount of voltage drop across the connecting wires and what is the voltage at the source?

Solution: The resistance of a conductor can be calculated from

$$R = \rho \frac{L}{A}$$

The specific resistance of 1 mil foot of copper wire is 10.4. The diameter of the conductor is 1/8 in. = 0.125 inch = 125 thousandths of an inch. The cross-sectional area in circular mils is $125 \times 125 = 15{,}625$ circular mils.

$$R = \frac{10.4 \times 300 \times 2}{15{,}625} = 0.39936 \ \Omega$$

The voltage drop across the two feed lines is $I \times R = 3 \times 0.39936 = 1.198$ V. The voltage at the motor is 110 V and the drop across the conducting lines is 1.198 V. The sum of these two is the source voltage: $110 + 1.198 = 111.198$ V.

Example:

The field coil of a DC shunt-wound motor has 3,400 turns of wire having a resistance of 80 Ω. The motor is connected to a source that supplies 44 V. What is the amount of power required by this winding?

Solution: The current flowing through this winding is

$$I = \frac{E}{R} = \frac{44}{80} = 0.55 \text{ A}$$

The power can be calculated from

$$P = E \times I = 44 \times 0.55 = 24.2 \text{ W}$$

It can also be determined by

$$P = \frac{E^2}{R} = \frac{44^2}{80} = \frac{44 \times 44}{80} = 24.2 \text{ W}$$

Example:

A wire that is 14 in. long moves at right angles to a uniform magnetic field whose flux density is 9×10^{-1} webers per square meter. What is the force in newtons that acts on this wire when a current of 25 A flows through the wire?

Solution:

$$F = B \times l \times I$$

The length of the wire must be expressed in the metric system: 1 in. = 2.54 cm; $14 \times 2.54 = 35.56$ cm = 0.3536 m.

$$F = 9 \times 10^{-1} \times 0.3536 \times 10^{-1} \times 25 = 7.96 \text{ newtons}$$

Example:

A certain load requires that the armature of a motor rotate at a speed of 2,000 rpm and supply a torque of 50 lb-ft. What is the horsepower output of this motor?

Solution:

$$hp = \frac{2\pi \times T \times N}{33,000}$$

$$= \frac{6.28 \times 50 \times 2000}{33,000} = 19 \text{ hp}$$

Example:

A DC motor having an input of 9 kW is required to supply 7 hp to its load. What is the efficiency of the motor under these working conditions?

Solution:

$$eff = \frac{\text{Output}}{\text{Input}}$$

$$= \frac{7 \times 746}{9000} = \frac{5222}{9000} = 0.58 = 58\%$$

Example:

The armature of a 120-V, 60-Hz AC motor runs at a synchronous speed of 3,600 rpm. How many poles does this motor have?

Solution:

$$P = \frac{F \times 120}{rpm} = \frac{60 \times 120}{3,600} = 2 \text{ poles}$$

Example:

What is the synchronous speed of an AC motor having a source voltage of 120 V, 60 Hz? The motor has a four-pole field winding.

Solved Motor Problems

Solution:

$$rpm = \frac{F \times 120}{P} = \frac{60 \times 120}{4}$$

$$= 1{,}800$$

Example:

What is the cross-sectional area of a copper wire whose radius is 1/4 in?

Solution: Convert inches to mils: 1/4 inch = 250 mils. The diameter = 2 × radius = 2 × 250 = 500 mils. The circular mil area = d^2 = 500 × 500 = 250,000 circular mils.

Example:

Three resistors having values of 4, 10, and 20 Ω are wired in parallel. What is the total resistance?

Solution: Since the component having the smallest resistance is 4 Ω, the total resistance must be less than this amount.

$$R_t = \frac{1}{\dfrac{1}{R1} + \dfrac{1}{R2} + \dfrac{1}{R3}}$$

$$= \frac{1}{\dfrac{1}{4} + \dfrac{1}{10} + \dfrac{1}{20}} = \frac{1}{0.25 + 0.1 + 0.05} = \frac{1}{0.4} = 2.5\ \Omega$$

Example:

A current of 4,750 mA flows through a pair of series field coils, each of which has a resistance of 1.035 Ω. What is the amount of electrical power in kilowatts delivered by the source?

Solution:

$$4{,}750\ \text{mA} = 4.75\text{A}$$

$$P = I^2R = (4.75)(4.75)(2.07) = 46.704\ \text{W} = 0.046\ \text{kW}$$

Note: Since the field coils are in series, their total resistance is twice that of a single one.

Example:

A battery has a terminal no-load voltage of 13.45 V and an internal resistance of 0.01365 Ω. A load connected across the battery draws a current of 945 mA. What is the load voltage of the battery?

Solution:

$$945\ \text{mA} = 0.945\ \text{A}$$

The internal voltage drop of the battery = $I \times R$ = 0.945 × 0.01365 = 0.129 V

$$\text{Load voltage} = E_{\text{no load}} - E_{\text{internal}} = 13.45 - 0.129 = 13.321\ \text{V}$$

Example:

What is the voltage across each of the resistors in Figure 8-1?

Solution:

$$120 \text{ mA} = 0.120 \text{ A}$$

$R2$ and $R3$ have identical resistances. This parallel group has a total value of 30 Ω. The same amount of current flows through each.

For $R3$,

$$E = I \times R = 0.12 \times 60 = 7.2 \text{ V}$$

For $R2$, $E = 7.2$ V since the resistors are in parallel. The total circuit current = 120 mA + 120 mA = 240 mA = 0.24 A. For R1,

$$E = I \times R$$
$$= 0.24 \times 100 = 24 \text{ V}$$

The voltage across $R4$, $R5$, and $R6$ equals the source voltage minus the sum of the voltage drops across $R1 + R2 = 24 + 7.2 = 31.2$ V:

$$40 \text{ V} - 31.2 \text{ V} = 8.8 \text{ V}$$

$$E_{R1} = 24 \text{ V};\ E_{R2,R3} = 7.2 \text{ V};\ E_{R4,R5,R6} = 8.8 \text{ V}$$

$$E_t = 24 + 7.2 + 8.8 = 40 \text{ V}$$

Example:

Determine the magnetic field intensity that exerts a force of 400 dynes on a magnet having a 50-unit pole strength.

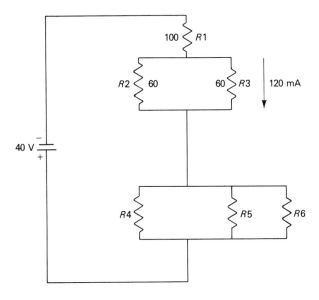

Figure 8-1

Solution:

$$f = m \times H$$

$$H = \frac{f}{m} = \frac{400}{50} = 8 \text{ oersteds}$$

Example:

A bar magnet has a flux of 10^5 maxwells. The rectangular end of the magnet measures 0.5×1 in. What is the flux density in gausses?

Solution:

$$1 \text{ in.} = 2.54 \text{ cm}$$

$$0.5 \text{ in.} = 2.54 \text{ cm}/2 = 1.27 \text{ cm}$$

$$A = 1.27 \times 2.54 = 3.23 \text{ cm}^2$$

$$B = \frac{\phi}{A} = \frac{10^5}{3.23} = 30{,}960 \text{ gauss}$$

Example:

What is the flux density in lines per square inch for a bar magnet whose end pole measures 2×3 in. and whose flux is 5,500 maxwells?

Solution:

$$A = 2 \times 3 = 6 \text{ in.}^2$$

$$B = \frac{\phi}{A} = \frac{5{,}500}{6} = 917 \text{ lines per square inch}$$

Example:

What is the reluctance of a section of carbon steel having a permeability of 2,500?

Solution:

$$\nu = \frac{1}{\mu} = \frac{1}{2{,}500} = 0.0004$$

Example:

The reluctivity of the magnet path of a field winding laminated core is 0.000412. The length of this magnetic path is 12 cm, and its cross-sectional area is 4 cm^2. What is the amount of reluctance?

Solution:

$$\mathcal{R} = \frac{\nu \times l}{A}$$

$$= \frac{0.000412 \times 12}{4} = 0.001236 \text{ rel}$$

Example:

A field coil has 180 turns and a magnetomotive force of 59.2 gilberts. How many ampere-turns are required?

Solution:

$$F = 1.26\, N \times I$$

$$N \times I = \frac{F}{1.26} = \frac{59.2}{1.26} = 47$$

Example:

An armature coil made of 60 turns of hard drawn copper wire is rated at 80 ampere-turns. How much current flows in this winding?

Solution:

$$N \times I = 80$$

$$I = \frac{80}{N} = \frac{80}{60} = 1.33\text{ A}$$

Example:

What is the amount of magnetomotive force in a magnetic circuit having a flux of 25,000 lines and a reluctance of 0.003147 rel?

Solution:

$$F = \phi \times \mathcal{R} = 25{,}000 \times 0.003147 = 79 \text{ gilberts}$$

Example:

What is the reluctance of a magnetic circuit if the reluctivity is 0.00035? The length of the magnetic path is 20 cm, and the area is 4 cm².

Solution:

$$\mathcal{R} = \frac{v \times l}{A} = \frac{0.00035 \times 20}{4} = \frac{0.007}{4} = 0.00175 \text{ rel}$$

Example:

The flux density of a magnet is 15,000 gauss and the magnetomotive force is 20 oersteds. What is its permeability? Its reluctance?

Solution:

$$\mu = \frac{B}{H} = \frac{15{,}000}{20} = 750$$

$$v = \frac{1}{\mu} = \frac{1}{750} = 0.00133 \text{ rel}$$

Example:

The laminations of an armature core consist of three parallel magnetic paths with reluctances of 0.0030, 0.0004, and 0.0025. What is the total reluctance?

Solved Motor Problems

Solution:

$$\mathcal{R}_t = \cfrac{1}{\cfrac{1}{\mathcal{R}1} + \cfrac{1}{\mathcal{R}2} + \cfrac{1}{\mathcal{R}3}}$$

$$= \cfrac{1}{\cfrac{1}{0.0030} + \cfrac{1}{0.0004} + \cfrac{1}{0.0025}}$$

$$= \cfrac{1}{333.33 + 2500 + 400}$$

$$= \cfrac{1}{3233.33} = 0.000309 \text{ rel}$$

Example:

The peak current having a sine waveform flowing through a motor dropping resistor is 850 mA. What is the average voltage developed across this resistor whose resistance is 65 Ω?

Solution:

$$850 \text{ mA} = 0.85 \text{ A}$$

$$E = I \times R = 0.85 \times 65 = 55.25 \text{ V (peak)}$$

$$E_{average} = 0.637 \times E_{peak}$$

$$= 0.637 \times 55.25 = 35.19 \text{ V average}$$

Example:

The average voltage of a sine wave measured across a resistor in a motor control circuit is 128 V. What is the amount of the peak voltage?

Solution:

$$E_{peak} = \frac{E_{average}}{0.637} = \frac{128}{0.637} = 200.94 \text{ V peak}$$

Example:

The average sine wave voltage measured across a resistor is 200 V. What is the peak-to-peak voltage?

Solution:

$$E_{peak} = 1.57 \times E_{average}$$

$$= 1.57 \times 200 = 314 \text{ V peak}$$

$$E_{peak\text{-}to\text{-}peak} = 2 \times E_{peak} = 2 \times 314 = 628 \text{ V}$$

Example:

A synchronous motor having two pairs of poles operates from a 120-V power line having an operating frequency of 60 Hz. What is the operating speed of this motor?

$$rpm = \frac{60 \times f}{p} = \frac{60 \times 60}{4} = 900$$

Example:

How much current, in milliamperes, flows through a 60-Ω resistor that dissipates 100 W?

Solution:

$$I = \sqrt{\frac{P}{R}} = \sqrt{\frac{100}{60}} = 1.29 \text{ A}$$

$$1.29 \text{ A} \times 1,000 = 1,290 \text{ mA}$$

Example:

A single-phase motor has an input power of 300 W. The AC voltage of the power line is 121 V, and the input current to the motor is 3.15 A. What is the power factor?

Solution:

$$pf = \frac{P}{E \times I} = \frac{300}{121 \times 3.15} = \frac{300}{381.15} = 0.787$$

$$= 78.7\% \text{ power factor}$$

Example:

A three-phase star-wound motor is connected to a 208-V line. It draws a current of 8 A and has a 60% power factor. How much input power is required?

Solution: Since the motor windings are inductive, the motor is a reactive load.

$$P = E \times I \times pf \times 1.732$$
$$= 208 \times 8 \times 0.60 \times 1.732 = 1729.223 \text{ W}$$
$$= 1.729 \text{ kW}$$

Example:

What is the current requirement of a 6-hp, three-phase motor whose power line input is 220 V? The efficiency of the motor is 85% and its power factor is 80%.

Solution:

$$I = \frac{746 \times hp}{E \times eff \times pf \times 1.732}$$

$$= \frac{746 \times 6}{220 \times 0.85 \times 0.80 \times 1.732}$$

$$= \frac{4476}{259} = 17.28 \text{ A}$$

Solved Motor Problems

Example:

What is the current requirement of a 5-hp, two-phase motor whose power line input is 208 V? The motor has an efficiency of 90% and its power factor is 80%.

Solution:

$$I = \frac{746 \times hp}{E \times eff \times pf \times 2}$$

$$= \frac{746 \times 5}{208 \times 0.90 \times 0.80 \times 2}$$

$$= \frac{3730}{299.5} = 12.45 \text{ A}$$

Example:

A DC motor operates from a 121-V AC line and requires 6.3 kW of electrical power. How much current does this motor draw?

Solution:

$$I = \frac{kilowatts \times 1{,}000}{E}$$

$$= \frac{6.3 \times 1{,}000}{121} = \frac{6{,}300}{121} = 52.0 \text{ A}$$

Example:

A current of 250 mA flows through a motor field coil having 400 turns. What is the magnetomotive force in ampere-turns? In gilberts?

Solution:

$$F = I \times N \text{ ampere turns}$$

convert 250 mA to amperes

$$250 \text{ mA} = 0.25 \text{ A}$$

$$F = 0.25 \times 400 = 100 \text{ ampere turns}$$

To change ampere turns to gilberts multiply by 0.4π or 1.257

$$F = 100 \times 1.257 = 125.7 \text{ gilberts}$$

Example:

What is the speed of a synchronous motor operating at a line frequency of 60 Hz? The motor is equipped with 4 poles.

Solution:

$$rpm = \frac{f \times 120}{P}$$

$$= \frac{60 \times 120}{4}$$

$$= 1800 \text{ rpm}$$

Example:

What is the percentage slip of a motor whose synchronous speed is 1800 rpm and whose operating speed is 1730 rpm.

Solution:

$$\text{percentage slip} = \frac{(\text{synchronous speed} - \text{operating speed})}{\text{synchronous speed}} \times 100$$

$$= \frac{1800 - 1730}{1800} \times 100 = 3.88\%$$

Example:

What is the full load current of a 5-hp 240-volt DC motor operated from a DC motor generator. The motor has an efficiency of 78% and has an output power of 4 hp.

Solution:

$$\text{Line current} = \frac{hp \times 746}{E \times eff}$$

$$= \frac{5 \times 746}{240 \times 0.78}$$

$$= \frac{3730}{187.2} = 19.93 \text{ A}$$

Example:

A motor rated at 5 hp has a power input of 4 kW. What are the power losses in this motor? What is its efficiency?

Solution:

$$5 \text{ hp} = 5 \times 746 = 3{,}730 \text{ watts}$$

$$4 \text{ kW} = 4{,}000 \text{ watts}$$

$$4{,}000 - 3{,}730 = 270 \text{ watts power losses}$$

$$\text{Efficiency} = \frac{\text{power output}}{\text{power input}} = \frac{3{,}730}{4{,}000} = 0.9325$$

$$0.9325 \times 100 = 93.25\% \text{ efficiency}$$

Solved Motor Problems

Example:

A 5-hp motor operating from a 120-volt source has an efficiency of 72%. Ignoring the resistance of the connecting power line, what is the amount of current being supplied to this motor?

Solution:

$$I = \frac{746 \times hp}{E \times \eta}$$

$$= \frac{746 \times 5}{120 \times 0.72} = \frac{3730}{86.4}$$

$$= 43.17 \text{A}$$

Example:

An electronic power supply delivers 65 volts to a small DC motor with this voltage measured when the motor is not running. After the motor has been turned on and is operating for 15 minutes the voltage is measured once again and is now 57 volts. What is the regulation of the power supply?

Solution:

$$\text{Percentage voltage regulation} = \frac{E_{no\ load} - E_{full\ load}}{E_{no\ load}} \times 100$$

$$= \frac{65 - 57}{65} \times 100 = 12.3\%$$

Example:

An electronic power supply develops a sine voltage having a peak-to-peak value of 185 volts. What is the average value?

Solution:

$$E_{average} = 0.637 \times E_{peak}$$

$$E_{peak} = \frac{E_{peak\text{-}to\text{-}peak}}{2}$$

$$= \frac{185.2}{2} = 92.5 \text{ V}$$

$$E_{average} = 0.637 \times 92.5 = 58.92 \text{ V}$$

Example:

The field coil of an AC motor having an input of 60 Hz has an inductance of 1,828 mH. What is the reactance of this coil?

Solution:

$$X_L = 2 \times \pi \times f \times L$$
$$= 2 \times 3.1416 \times 60 \times 1.828$$

Note: 1,828 mH = 1.828 H

$$= 689.14 \text{ ohms}$$

Example:

A three-phase motor used to operate a machine lathe is connected to an input of 440 volts, 60 Hz, with an operating current of 4 amperes. How is this motor rated in terms of kilovolt-amperes?

Solution:

$$\text{KVA} = \frac{E \times I \times 1.732}{1{,}000}$$
$$= \frac{440 \times 4 \times 1.732}{1{,}000} = \frac{3048.32}{1{,}000} = 3.048 \text{ KVA}$$

Index

A

AC:
   average values, 104
   magnetic field, 146
   for motors, 101
   Ohm's Law, 116
   sine wave, 103
   voltage and current measurements, 103
   waveforms, 102
AC motors, 209
   basic, 209
   direction of rotation, 257
   types of, 210
Alkaline cells, 78
Amortisseur winding, 218
Ampere:
   -hour ratings, 70
   -turn, 155
Armature, 168
   amortisseur, 218
   connections, non-symmetrical, 179
   connections, symmetrical, 179
   force, 181
   poles, 180
   rpm, 229
   squirrel cage, 218
   wave windings, 180

Armature coil:
   back pitch, 175
   front pitch, 175
   magnetic field, 170
   number of turns, 169
   pitch, 175
Armature winding, 171
   duplex, 177
   lap, 173
   progressive, 175
   retrogressive, 176
   simplex, 176
   simplex progressive lap, 177
   total lap, 178
   triplex, 177
   types, 172
Autotransformer speed control, 251
Average values of AC, 104

B

Back pitch, 175
Basic data, 1
Batteries, 67
   alkaline, 78
   characteristics of alkaline, 79
   characteristics of lead-acid, 74

Batteries *(cont.)*
  charge rate, 87
  dimensions of alkaline, 79
  efficiency, 85
  identification, 72
  internal resistance, 84
  lead-acid, 73
  lithium, 81
  lithium characteristics, 82
  lithium dimensions, 83
  mercury, 78
  mercury sizes, 81
  mercury types, 80
  nickel-cadmium, 74
  nickel-cadmium charging rate, 76
  physical dimensions, 67
  silver oxide, 82
  silver oxide dimensions, 83
  types and characteristics, 71
  zinc-air, 84
  zinc-carbon, 75
  zinc-carbon characteristics, 77
Bearing oil, 187
Bearings, motor, 187
Braking, 264
Brush:
  holders, 167
  materials, 164, 166
  selection, 167
Brushes, 164
  electro-graphitic, 165
  electrolytic action, 167
  graphite, 167
  hard carbon, 164
  metal graphite, 167
Brushless dc motors, 205
  characteristics, 206

C

Capacitive reactance, 115
Capacitor:
  motor, single-phase, 258
  split-phase motors, 230
  split-phase motors, characteristics, 231
  -start two-speed motor, 232
Cartridge fuse, 50
Cell polarity, 70
Cells:
  in parallel, 69
  in series, 68
  in series opposing, 70
  in series-parallel, 69
Celsius-fahrenheit conversions, 40
Centrifugal switches, 236
CGS systems, 136
Chopper, 256

Circuit breakers, 51
Coercive force, 159
Coil pitch, 175
Coils, field, 183
Commutator, 163
  cleaning, 164
  pitch, 175
Compound:
  dc motor, differentially wound, 195
  motor, cumulative-wound dc, 196
  -wound dc motor, characteristics, 195
  -wound motor, 194
Conductivity, 21
Conduit box, 186
Control circuits, electric/electronic motor, 250
Conversion factors for MKS, CGS and English
    units, 138
Conversions, voltage, 2, 3
Core-loss aging coefficient, 158
Cumulative-wound compound motor, 196
Curie point, 158
Current, 5
  conversions, 6
  dc motor full-load, 58
  designations, 7
  versus electrical power input, 223
  heating effect, 35
  and power, 118
  requirements of polyphase motors, 242
  requirements for single-phase motors, 222
  starting versus running, 228
  in three-phase circuits, 130, 246
  transformers, step up and step down, 134
  versus wire sizes, 47
  and voltage, effective values, 106
  and voltage, instantaneous values, 105
  and voltage measurements, 103
  and voltage phase relationships, 120

D

Data plate, 199
DC:
  circuits, power, 27
  compound motor, cumulative-wound, 196
  compound wound motor, reversing, 261
  magnetic field, 146
  power laws, 28
  power sources, 67
  shunt-wound motor, reversing, 261
DC motors, 162
  brushless, 205
  characteristics of brushless, 206
  characteristics of compound-wound, 195
  compound-wound, 194
  full-load currents, 58
  series-wound, characteristics, 189
  shunt-wound, 190

Index

split-field series-wound, 190
types, 163
Density, flux, 147
Diac, 255
and triac speed control, 265
Diamagnetic materials, 157
Differentially wound compound motor, 195
Diode voltage control, 92
Disk stepper, 200
Drip-proof frame, 185
Dual polarity power supplies, 87
Duplex winding, 177
Dust-proof frame, 186
Dynamotor, 98

E

E, I, and R conversions summary, 12
Eddy current power loss, 183
Effective values of sine voltages and currents, 106
Efficiency, 62, 131, 246
horsepower and power factor in three-phase motors, 246
Electric/electronic motor control circuits, 250
Electric versus magnetic circuits, 154
Electronic power supplies, 86, 250
Enclosed frame, 185
Enclosure, 184
Energy versus power, 35, 117
English:
and metric units, 58
systems, 136
to metric, 141
units, 138
Explosion-proof frame, 186

F

Fahrenheit, Celsius conversions, 40
Fan-cooled frame, 186
Feet to meters, 61
Ferromagnetic materials, 157
Field:
coils, 183
current in shunt motor, 192
Filter, RFI, 266
Flux:
density, 147
lines, 140
lines, characteristics, 143
Foot-pound, 56
Force:
on an armature, 181
coercive, 159
on a conductor, 145
Fractional-horsepower:

motors, 260
reversing motors, 262
Frames:
cooling bands, 186
drip-proof, 185
dust-proof, 186
explosion-proof, 186
heat dissipation, 186
internal fan cooled, 186
motor, 184
mounts, 187
open, 184, 186
protected, 185
splash proof, 185
totally enclosed, 185
variations, 186
vented, 186
watertight, 186
Frequency:
generator, 112
power-line, 220
Front pitch, 175
Fuse:
cartridge, 50
current ratings for repulsion-induction motors, 216
pullers, 50
ratings, 48
types, 47
Fuses for motors, 47
Fusing currents of wires, 46

G

Generator:
frequency, 112
types, 112

H

High-frequency motor control, 274
Horsepower, 53
to amperes conversion, 244
efficiency and power factor in three-phase motors, 246
versus watts, 55
Hybrid stepping motor, 203
Hysteresis, 160
loss, 161

I

Impedance, 115
Incremental mode, 200
Induction:
motor control, 268

Induction *(cont.)*
  motors, polyphase, 240
  residual, 159
Inductive reactance, 113
Input power versus output power, 128
Instantaneous values of sine voltages and currents, 105
Insulation, 204
Insulators, 23
Interpoles, 184
IR drops, 25
  polarity, 25

K

Kilohm, 8
Kilovolt, 3
KVA:
  single phase, 126
  three-phase, 126
  and VA, 125

L

Lap winding, 173
Lead-acid batteries:
  characteristics, 74
  charging rate, 74
Linear measurements in English and metric, 140
Lines, flux, 140
Lithium cells, 81
Loading, 27
Loss:
  hysteresis, 161
  power line, 28, 220
Low-speed voltage regulator, 270

M

Magnetic:
  aging, 158
  circuits, formulas, 136
  versus electric circuits, 154
  field, ac, 146
  field, armature, 170
  field, dc, 146
  force around a conductor, 144
  force between poles, 145
  formulas, manipulating, 155
  and non-magnetic materials, 157
  saturation, 158
  substances, reluctivity, 153
Magnetic lines, 140
  characteristics, 143
  shape, 147

Magnetomotive force, 154
Maxwell, 143
Mechanical and electrical power equivalents, 57
Megavolt, 4
Megohm, 8
Mercury cells, 78
Meters to feet, 60
Metric:
  and English units, 58, 141
  units, 138
Micro:
  ampere, 6
  motors, 277
  steps, 200
  volt, 3
Milli:
  ampere, 5
  volt, 3
MKS system, 136
MKSA system, 137
Motors:
  ac, 209
  basic ac, 209
  bearings, 187
  braking, 264
  brushless dc, 205
  capacitor split-phase, 230
  capacitor-start two-speed, 232
  compound-wound, 194
  control circuits, electric/electronic, 250
  control, induction, 268
  cumulative-wound, compound dc, 196
  current for polyphase, 242
  current for single-phase, 222
  Darlington control, 95
  dc, 162
  differentially wound compound dc, 195
  direction of rotation of ac, 257
  distributed field compensated, 214
  field current in shunt dc, 192
  fractional-horsepower, 260
  frames, 184
  fuse current ratings for repulsion induction, 216
  high-frequency control, 274
  horsepower, efficiency, and power factor, 246
  hybrid stepping, 203
  low-voltage control, 92
  operating voltage, 193
  permanent-capacitor, 233
  plug and socket symbols, 214
  PM stepping, 203
  polyphase, 238
  polyphase induction, 240
  polyphase induction, reversing, 261
  power delivered to three-phase, 248
  repulsion, 215

Index

repulsion-induction, 216
repulsion-start induction, 215
reversing circuitry, 269
reversing fractional hp, 262
self-starting induction-reaction synchronous, 226
series-wound, 188
series-wound dc, 189
series-wound reversing, 261
shaded-pole induction, 225
shafts, 187
shunt-wound dc, 190
single-phase, 219
single-phase and polyphase, 214
single-phase capacitor, 258
single-phase series, 259
solved problems, 278
speed control, 188
speed control of battery-operated, 256
speed control, shunt-wound dc, 267
speed control, universal, 213, 257
speed and regulation control, 263
split-capacitor, 235
split-field series-wound dc, 190
split-phase, 227, 231, 236, 259
squirrel-cage, 219
stepless, 199
stepping, 199
symbols, 205
synchronous, 226
tachometer control, 271
terminal connections for 110 V/220 V, 249
terminology of stepper, 204
three-phase, 248
two-capacitor, 233
two-phase, 259
two-phase, operating from a three-wire line, 248
types of, ac, 210
universal, 211
variable reluctance, 201
winding, polyphase, 238
wound-rotor induction, 240

N

Name plate, 199
NI, 155
Nickel-cadmium cells, 74
Nomograms, 35
Non-symmetrical connections, armature, 179

O

Ohm's Law:
  ac, 116
  dc, 24
  for magnetic circuits, 141
Oil, bearing, 187
Open frame, 184
Output power versus input power, 128

P

Paramagnetic materials, 157
Permanent:
  -capacitor motor, 233
  -magnet rotor, 201
Permeability, 150
  relative, 152
  versus reluctivity, 152
Permeance, 153
PF, 116
Phase, 110
  KVA three-, 126
  relationships, current, and voltage, 120
  single, 119
  single KVA, 126
  three, 129
  two, 133
Pitch:
  back, 175
  coil, 175
  commutator, 175
  front, 175
Plug:
  -to circuit breaker conversions, 52
  fuse, 49
  and socket symbols, 214
PM:
  rotor, 201
  stepping motor, 203
Poles, armature, 180
Polyphase:
  induction motors, 240
  induction motors, reversing, 261
  motor windings, 238
  motors, 238
  motors, current requirements, 242
  and single-phase motors, 214
Pound-feet, 60
Power, 248
  conversions, 29, 59
  and current, 118
  in dc circuits, 27
  delivered to a three-phase motor, 248
  designations, resistor, 9
  dissipation, total, 32
  versus energy, 35, 117
  factor, 116
  input versus output, 128
  laws, dc, 28
  loss, eddy current, 183
  real versus reactive, 124

Power *(cont.)*
  relationships, 32
  sources, dc, 67
  in three-phase circuits, 129
  three-phase delta-wye, 252
  transmission, 198
Power-line:
  frequency, 220
  losses, 28, 220
Power supplies:
  dual polarity, 87
  electronic, 86, 250
Practical system, 138
Progressive lap winding, simplex, 177
Protected frame, 185

R

Reactance, 113
  capacitive, 115
  inductive, 113
Reactive power versus real power, 124
Regulation control, 263
Relays, 65
Reluctances, 148
  in parallel, 150
  in series, 149
Reluctivity:
  of magnetic substances, 153
  versus permeability, 152
Remanence, 159
Repulsion:
  -induction motors, characteristics, 216
  motors, 215
  motors, characteristics, 215
  -start induction motors, 215
  -start induction motors, characteristics, 215
Residual induction, 159
Resistance, 8
  conversions, 8
  conversions in English and metric, 59
Resistivity of metals, 21
Resistors:
  characteristics, 10, 11
  codes and values, 12
  color code, 12
  in parallel, 15, 16
  power designations, 9
  in series, 15
  in series-parallel, 17
  symbols, 12
  tolerance, 14
  types, 9
Retentivity, 159
Retrogressive winding, 176
RFI filter, 266
Right-hand motor rule, 162
Rotary converter, 99

Rotation of ac motors, direction, 257
Rotor:
  permanent-magnet, 201
  slip, 241
  squirrel-cage, 217
Rpm, armature, 229

S

Saturation, magnetic, 158
SCR speed control, 255
Second, standard, 137
Self-starting:
  induction-reaction synchronous motor, 226
  techniques for split phase, 228
Sensor-servo control, 276
Series motor:
  single-phase, 259
  speed control, 275
Series voltage drops, 27
Series-wound:
  dc motor, characteristics, 189
  motor, 188
  motor, reversing, 261
  split-field, 190
Shaded-pole:
  induction motor, characteristics, 225
  motor, 224
Shaft, motor, 187
Shunt:
  motor, field current in dc, 192
  -wound dc motor speed control, 267
  -wound motor, 190
  -wound motor, reversing the dc, 261
SI units, 137
Silicon-controlled rectifier, 255
Silver-oxide cells, 82
Simplex:
  progressive lap winding, 177
  winding, 176
Sine wave, 103
  relationships, 109
Single phase, 119
  capacitor motor, 258
  motors, 219
  and polyphase motors, 214
  series motor, 259
Slip, rotor, 241
Slotless stator, 207
Socket and plug symbols, 214
Source voltage, 4
Speed control, 188, 263
  autotransformer, 251
  of battery-operated series-wound dc motors, 256
  diac and triac, 265
  SCR, 255
  series motor, 275

Index

universal motor, 257
of universal motor, 213
Speed, synchronous, 240
Splash-proof frame, 185
Split-capacitor motor, 235
  summary of characteristics, 235
Split-field series-wound motor, 190
Split-phase motors, 227, 259
  summary of characteristics, 236
Squirrel-cage:
  concept, 217
  double, 218
  motors, characteristics of, 219
Standard second, 137
Starting versus running current, 228
Stator, slotless, 207
Step:
  angles versus steps, 200
  frequency, 200
  versus step angle, 200
Stepless motors, 199
Stepper:
  disk, 200
  motors, terminology, 204
Stepping motors, 199
  PM, 203
  types, 199
Surge limiting, 90
Switches, centrifugal, 236
Symbols:
  motor, 205
  socket and plug, 214
Symmetrical connections, armature, 179
Synchronous:
  motor, 205
  speeds, 227, 240

T

Tachometer motor control, 271
Temperature, 38
  coefficient of resistance, 41
  conversions, 39
Terminal connections for 110-V/220-V motors, 249
Terminology of stepper motors, 204
Three phase, 129
  circuits, current, 130, 246
  circuits, power, 129
  delta-wye power, 252
  motor, characteristics, 248
  rectifier systems, 253
Torque, 60
  control, 269
Total lap winding, 178
Transformers:
  current, step up and step down, 134
  voltage, step up and step down, 134

Transient voltages, eliminating, 266
Transistor voltage regulators, 93
Transmission of power, 198
Triac, 255
  and diac speed control, 265
Triplex winding, 177
Two-phase, 133
  motors operating from a three-wire line, 248
  motors, 259

U

Universal motor, 211
  characteristics, 212
  distributed field compensated, 214
  speed control, 213, 257

V

VA and KVA, 125
Variable reluctance motors, 201
VARS, 126
Vented frame, 186
Voltage, 1
  control, diode, 92
  control, shunt transistor, 94
  conversions, 2
  -current, effective values, 106
  -current, instantaneous values, 105
  -current measurements, 103
  -current phase relationships, 120
  current and resistance relationships, 23
  drops, series, 27
  eliminating transient, 266
  motor operating, 193
  reference points, 71
  regulation with differential amplifier, 94
  regulation with series Zener and transistor, 94
  regulators, 89, 93, 270
  source, 4
  transformers, step up and step down, 134

W

Watertight frame, 186
Watts:
  versus horsepower, 54
  multiples, 29
Wave windings, 180
Waveforms, ac, 101
Wheatstone bridge control, 270
Windings:
  duplex armature, 177
  lap, 173
  progressive armature, 175

Windings *(cont.)*
   retrogressive armature, 176
   simplex, 176
   simplex progressive lap, 177
   total lap, 178
   triplex armature, 177
   types of armature, 172
   wave, 180
Wire, 42
   cross-sectional area, 44
   fusing currents, 46
   resistance, 44
   table, 43

Work, 33
Wound-rotor induction motors, 240

Z

Zener:
   -diode regulator, 90
   regulation, cascaded, 92
Zeners, series, 91
Zinc:
   -air cells, 84
   -carbon cells, 75
   -chloride cells, 83